Best Ma

Integrated Vegetation Management

Third Edition

Randall H. Miller

Produced in collaboration with the Utility Arborist Association

UAA
UTILITY ARBORIST ASSOCIATION

Companion publication to the *American National Standard—Tree, Shrub, and Other Woody Plant Management—Standard Practices (Integrated Vegetation Management)* (ANSI A300, Part 7)

ISA
International Society of
Arboriculture

ISBN: 978-1-943378-20-3

Printed in the United States of America.

Cover Design: Shawna Armstrong
Cover Photo: Joshua Salas/CNUC
Page Design and Composition: Veridian Design Group
Illustrations: Bryan Kotwica (all figures except Figure 2)
Figure 2: Quantum Spatial/NV5 GeoSpatial
Printed by Premier Print Group, Champaign, IL

International Society of Arboriculture
270 Peachtree St. NW, Suite 1900
Atlanta, GA 30303

+1.678.367.0981
www.isa-arbor.com
permissions@isa-arbor.com

10 9 8 7 6 5 4 3
0622-CA-2000

Contents

List of Figures

List of Tables

Acknowledgments

Review Committee

Peter Beesley, Pacific Gas and Electric

Josiane Bonneau, Wildlife Habitat Council

Phillip Charlton, PhD, Utility Arborist Association

John Goodfellow, BioCompliance Consulting, Inc.

Geoff Kempter, Asplundh Tree Expert, LLC

Matt Kraushar, Indiana Department of Transportation

Lew Payne, New York Power Authority

Anand Persad, PhD, ACRT

William Rees, Baltimore Gas and Electric (retired), W TREES Utility Vegetation Management Consulting, LLC

Brian Smith, United States Department of Transportation Federal Highway Administration

Becky Spach, FirstEnergy

Stan Vera-Art, Grow With Trees Company

Jamie Vidich, International Society of Arboriculture

Contributors

The following contributed advice or reviews to the project.

Bob Bell, Pacific Gas and Electric

Victoria Brinkly, United States Department of Transportation, Federal Highway Administration

Ed Frantz, New York State Department of Transportation

Craig Kelly, Pacific Gas and Electric

Dave Krause, Asplundh Tree Expert, LLC (retired)

Joe Lentz, Arborchem (retired)

Amanda Maxwell, Natural Resource Defense Council

Kevin McLoughlin, New York Power Authority (retired)

Sydney Mucha, North Carolina Department of Agriculture and Consumer Services (formerly with the Wildlife Habitat Council)

Larry O'Donnell, United States Department of Transportation, Federal Highway Administration

Tom Smiley, PhD, Bartlett Tree Experts

Robert Vanderhoof, PhD, Tall Tree Learning

Purpose

The purpose of this publication is to complement the *American National Standard—Tree, Shrub, and Other Woody Plant Management—Standard Practices (Integrated Vegetation Management)* (ANSI A300, Part 7) and provide practitioners with what industry experts consider to be the most appropriate integrated vegetation management (IVM) techniques.

In the United States in 1991, arboricultural professionals established a committee to improve professionalism and the quality of service to the public by developing American National Standards Institute (ANSI) standards for arboricultural operations. Since then, that committee, working under the authority of ANSI, developed standards for *Pruning, Soil Management, Supplemental Support Systems, Lightning Protection Systems, Construction Management, Planting and Transplanting, Integrated Vegetation Management, Root Management, Tree Risk Assessment*, and *Integrated Pest Management*. ANSI A300 standards for *Tree, Shrub, and Other Woody Plant Management* provide guidance for professional arborists in the development of work practices, maintenance specifications, best practices, regulations, and other measures of performance.

Abbreviations and Acronyms

AHAS: Acetohydroxyacid synthase

ALS: Acetolactate synthase

ANSI: American National Standards Institute

BMP: Best management practice

EPSP: 5-enolpyruvylshikimate-3-phosphate

ESG: Environmental, social, and governance

FERC: Federal Energy Regulatory Commission

GIS: Geographic information system

IA: Independent assurance

IPM: Integrated pest management

ISA: International Society of Arboriculture

IVM: Integrated vegetation management

Lidar: Light detection and ranging

MOA: Mode of action

MVCD: Minimum vegetation clearance distance

NEPA: National Environmental Policy Act

NERC: North American Electric Reliability Corporation

pH: Potential of hydrogen

QA: Quality assurance

QC: Quality control

RHWG: Rights-of-Way as Habitat Working Group

ROW: Right-of-way

SOA: Site of action

TGR: Tree growth regulator

UAA: Utility Arborist Association

WHC: Wildlife Habitat Council

1. Introduction

Integrated vegetation management (IVM) is based on the principles of integrated pest management (IPM). The primary goal of IPM is balanced use of control methods to maintain pest populations below tolerance levels. The "pest" in the case of IVM is vegetation incompatible with management goals and maintenance objectives.

IVM best practices can be applied where they are necessary to create and maintain desired vegetation characteristics and to facilitate the purpose and use of the land, notably utility and transportation rights-of-way (ROWs). In addition to restoring ecosystems, promoting pollinator forage, establishing early successional plant communities, improving early successional wildlife habitat, controlling invasive plants, and preserving cultural resources, IVM can be useful in arboricultural and landscape applications. It is an adaptive, interdisciplinary system that considers current scientific and stakeholder perspectives; provides a flexible structure to facilitate a comprehensive approach to vegetation management; strategically promotes and conserves sustainable, compatible vegetative cover types; and suppresses incompatible species.

Natural systems are dynamic, and managing them requires experience and knowledge of ecosystems and how they work. This publication is not intended to substitute for the expertise of a vegetation manager. A vegetation manager is an individual engaged in the profession of vegetation management who, through education and related training, has the competence to direct an IVM program. The expertise of a vegetation manager contrasts with that of an arborist insofar as vegetation managers focus on ecosystems, while arborists concentrate on individual trees or small groups of trees. Vegetation managers are responsible for setting goals and objectives; evaluating site conditions; establishing specifications; making decisions on tolerance levels, action thresholds, and control methods; and performing quality assurance. They should use applicable standards and this BMP to write or review specifications, statements of work, and contracts.

Practitioners are people employed internally or contracted by the organization with responsibility over IVM. Practitioners must have the necessary education and training to manage the complexities of IVM. Appropriate knowledge can be obtained through formal education, experience, or a combination of the two. Subject matter expertise should include ecosystem dynamics and how they apply to management areas, including a recognition of compatible and incompatible plant communities and how and when control methods are appropriate to achieve objectives. Competency is also necessary among workers. Training to improve and refine the knowledge of workers should be ongoing and up to date. Employment stability, training, and opportunities for advancement among the workforce are hallmarks of successful IVM programs.

Trees and other plants are living organisms, and they and the ecosystems in which they live are variable by nature, so not all practices can be successfully applied in every case, and many instances require multiple control methods. Departures from best management practices should only be made in response to conspicuous need with supporting rationale.

IVM Definition

The American National Standards Institute (ANSI A300, Part 7) defines IVM as a system of managing plant communities in which compatible and incompatible vegetation are identified; action thresholds are determined; tolerance levels are established; and control methods are evaluated, selected, and applied to achieve management goals and maintenance objectives. Integrated vegetation management uses a variety of controls and often integrates multiple methods to promote sustainable plant communities that are compatible with management goals. Incompatible vegetation compromises program goals, negatively impacting or causing concerns regarding safety, security, access, fire risk, utility service reliability, wildlife habitat, pollinator forage, emergency restoration, visibility, line-of-sight requirements, regulatory compliance, the environment, and much more. Integrated vegetation management is not a set of inflexible prescriptions, such as repeated mowing or broadcast spraying across entire areas on rigid schedules. Rather, it is a long-term process tailored to the species being managed, existing site conditions, and intended outcomes.

2. Implementation

Integrated vegetation management must comply with applicable laws and regulations and be in accordance with relevant standards and best practices. Notable regulations in the United States include the Endangered Species Act; the National Environmental Policy Act (NEPA); the Federal Insecticide, Fungicide, and Rodenticide Act; the Migratory Bird Treaty Act; the Clean Water Act; labor laws; and others. Standards include the North American Reliability Corporation *Vegetation Management Standard* (FAC-003-4), the *American National Standard for Arboricultural Operations—Safety Requirements* (ANSI Z133), the *American National Standard for Tree Care Operations—Tree, Shrub, and Other Woody Plant Management—Standard Practices* (ANSI A300), and the accompanying BMP series from the International Society of Arboriculture (ISA), including this publication.

Organizations that perform vegetation management should conduct themselves ethically, which means going beyond merely complying with laws. Reputations are built on integrity and transparency, so vegetation management operations should develop a culture of honesty that pervades their organization. This includes being prompt for appointments, returning phone calls and emails, and following up on commitments.

Safety

Without rigorous training and strict adherence to applicable safety procedures and regulations, vegetation maintenance operations can pose significant safety risks. For that reason, vegetation managers need to develop a culture of safety throughout their organizations beginning with employing qualified professionals who have demonstrated ability to work according to accepted safe practices or qualified trainees dedicated to learning those safe work practices. Workers should have the training and proper tools to safely complete the work to specifications. In the United States, the *American National Standard for Arboricultural Operations* (ANSI Z133) provides safety requirements for arboricultural work.

When properly performed, IVM also protects the public and other stakeholders. Examples of IVM safety goals are reducing vegetation density for access, protecting power lines from interference, reducing fire risk, enhancing roadway safety, protecting railway integrity, maintaining lines of sight, minimizing exposure to tree-related risk, and other factors. Specifics on electric utility, pipeline, railroad, and roadway IVM can be found in the appendixes.

Right-of-Use

Integrated vegetation management requires a long-term right to use the land for the intended purpose. Long-term right-of-use can include ownership of the property, franchise or prescriptive rights, leases, easements, or permits. To be successful, the long-term right to use IVM must be established on lands, sites, and facilities on which IVM is to be applied, including the right to access the management area for inspection and maintenance. Easements for access roads are often necessary and should be maintained with written plans for their periodic inspection and upkeep. Managers should ensure there are no potentially conflicting usage rights, such as conservation easements, water rights, native rights, organic farms, or other potential impediments to IVM.

Unauthorized uses of management areas should be discouraged. Alternative uses may be authorized with advance permission or agreement. Consequently, vegetation managers need to develop written policies that characterize approved uses according to rights granted on the site and should detail procedures for minimizing unauthorized uses. Asset managers should document wrongful uses that come to their attention and report those activities to the competent authorities.

3. IVM Process

Integrated vegetation management is an iterative process that comprises a vegetation management program and one or more vegetation maintenance plans. Vegetation management programs are strategic and long term, while vegetation maintenance plans are tactical and short term, more at a project level. Plans are adjusted based on variations in environmental and site conditions. The IVM process coordinates vegetation management program and maintenance plans in a documented procedure designed in a continuous loop (Figure 1). The process is cyclical because successfully managing natural systems must be ongoing and centered around adaptive management. The approach provides flexibility to adjust plans as circumstances evolve and new information becomes available.

Vegetation Management Programs

Vegetation management programs are formal and strategic with long-term horizons. They provide policies, procedures, goals, specifications, and

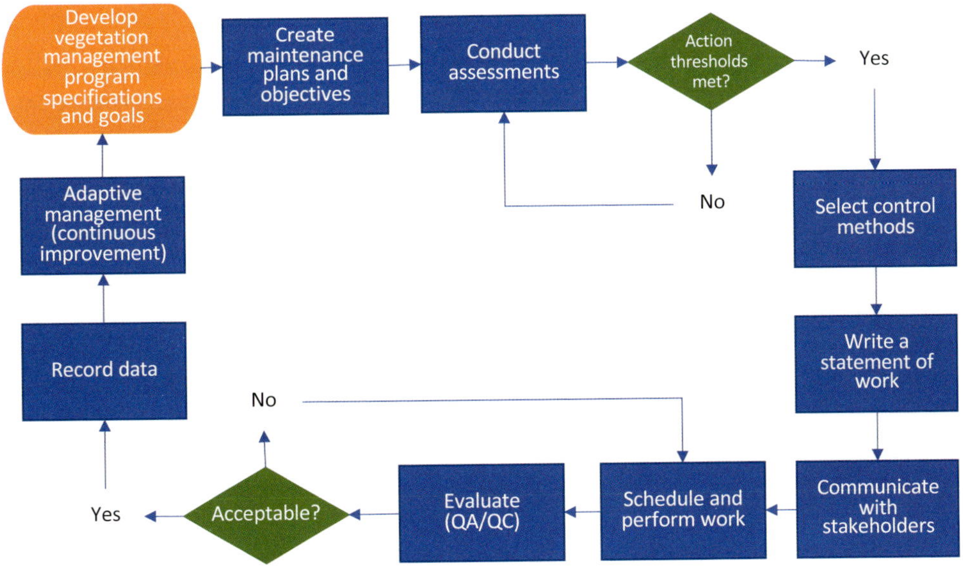

Figure 1. IVM process flow chart. Order of processes may vary or occur concurrently.

methodologies. Programs should be developed with the cooperation and involvement of appropriate external and internal stakeholders and should include vision, mission, and value statements with an overview. The overview should comprise a general characterization of available resources emphasizing the area under management, infrastructure attributes, and a land-use description. Strategic plans need to be revised periodically based on lessons learned during maintenance implementation (Figure 1).

Vegetation management goals are strategic outcomes for IVM programs that govern maintenance-level objectives. Examples are presented in Table 1.

Specifications

Specifications are detailed requirements, procedures, and standards used to define and govern actions, including writing statements of work. They provide guidance on applying strategic plans to short-term maintenance objectives. Specifications based on applicable standards and best practices have credibility and authority. They should be written by a vegetation manager in consultation with appropriate stakeholders. They must be written carefully because they may become part of legally binding contracts, and well-written specifications minimize misunderstandings that could otherwise result in conflicts and undesirable outcomes.

Specifications should be written to promote program goals. Among other considerations, they should include control methods and describe action thresholds and tolerance levels. Examples of possible specification elements are presented in Table 2.

Table 1. Examples of vegetation management goals.

- Accommodate the intended use of the site (i.e., transportation, environmental stewardship, or delivery of electricity or gas)
- Advance environmental stewardship and sustainability, including restoring or enhancing ecological benefits
- Comply with applicable laws and regulations
- Control incompatible species
- Ensure operational flexibility
- Maintain site security
- Optimize maintenance cost
- Promote public and worker safety
- Protect cultural sites
- Protect, enhance, and propagate compatible species

Table 2. Examples of possible specification elements.

• Control methods • Communication with stakeholders – Internal – External • Facility inspection • Locations of treatment sites	• Quality assurance • Reporting • Safety – Safe work rules – Fire prevention • Scheduling protocol • Storm-response procedures	• Tolerance levels and action thresholds – Compatible and incompatible species – Vegetation heights or clearances – Numbers or density of compatible and incompatible plant species

Determine Tolerance Levels and Action Thresholds

Tolerance levels and action thresholds are concepts adapted from IPM. In IPM, a tolerance level is the point at which pest pressures reach levels where there is an unacceptable economic threat. In IVM, tolerance levels are maximum incompatible plant pressures allowable without unacceptable consequences. Plant pressures are factors such as vegetation height, location, density, incompatibility, condition, or a blend of factors. Unacceptable consequences may include safety incidents, compliance violations, economic harm, environmental degradation, or a combination of these or other adverse outcomes. Tolerance levels must not be breached.

Action thresholds in IPM are levels of pest pressure where work should begin to prevent a critical economic threat. Action thresholds in IVM are levels of incompatible plant pressure where control measures should be implemented to prevent tolerance levels from being infringed.

Action thresholds and tolerance levels are linked insofar as action thresholds should be set to allow a margin of error before unacceptable conditions threaten tolerance levels. As conditions approach tolerance levels, risk increases dramatically. So, action thresholds are determined by optimizing the timing and cost of IVM treatments against the risk of unacceptable consequences. For example, action thresholds for electric transmission lines must be established at distances substantially farther from lines than the corresponding tolerance level (such as maximum flashover distances between lines of a specified voltage and vegetation). This allows work to be triggered long before unacceptable consequences such as service reliability, violation of regulatory requirements, or unsuitable safety risks develop. Best management practices establish action thresholds at a point where they will enable the program to maintain a stable, compatible plant community comprised of vegetation types that will not interfere with program goals, let alone encroach on tolerance levels. Because they vary from program to program, tolerance levels and action thresholds should be developed by vegetation managers and should include lists of incompatible plants or cover types.

Action thresholds should not be based on strict time intervals between control-measure applications. Variability in growth rates, species composition, environmental factors, past maintenance practices, facility use, land development, and other considerations will often cause work-timing requirements to fluctuate. For example, an overly long maintenance interval may result in a more intrusive intervention, which can negatively affect the site and create more work, expense, and environmental disturbance than would occur with more suitable scheduling; too short an interval might also create unnecessary expense. So, rather than set times, optimal maintenance intervals should be determined by focusing limited resources on areas of greatest need based on evaluations of site and workload.

Communication and Stakeholder Engagement

Communication and stakeholder engagement is a strategy for public relations and education. It is used to inform external and internal stakeholders regarding IVM. Communication is essential to planning and implementing a successful vegetation management program. Records of significant communication and events, including meeting minutes, are indispensable.

Proactive organizations anticipate stakeholder interests and provide information in a variety of formats. Concerns should be approached as an opportunity to favorably represent the vegetation management program. Honest, transparent information will build trust and serve in the long-term best interest of the operation.

External Stakeholder Communication

External stakeholders are groups and individuals outside of an IVM organization who will be affected by IVM activities. Examples include public land managers, property owners, regulators, nonprofits, governmental agencies, community partners, special interest groups, suppliers, and other parties that have justifiable concerns. The effect the work may have on stakeholders should not be dismissed or misrepresented. When dissatisfaction has merit, it should be acknowledged and the matter should be addressed promptly and with integrity. Not every issue can be resolved to stakeholder satisfaction, however. Those that cannot may be straightened out through actively listening and explaining how the planned work benefits the concerned party. While those involved might not agree with the reasons their needs cannot be met, at least they will understand they were heard. Appropriate communication should ensure concerns do not increase in frequency or severity over time.

Stakeholders can be met in person through public meetings or engaged through social media and electronic or printed notices. Examples of printed materials include mailed letters, brochures, pamphlets, door hangers, or public notices. Printed information should be well illustrated and easy to understand. Notification should emphasize the reason for the project and the benefits it provides. It should include the location, the timing of planned activities, potential crew types, and other information that might be helpful.

Work on governmentally managed property can involve administrative procedures that take months of advance work, including navigating through permit processes and the concerns of specialists who have responsibility for stewardship over public lands. Vegetation managers should strive to understand the responsibilities of land specialists and work with them to balance stewardship with other objectives. An example of government–private sector cooperation in the United States is a memorandum of understanding reached between Edison Electric Institute member utilities and federal land management agencies. The memorandum identifies common interests and promotes cooperation among the parties to ensure safe, reliable electric supply in an environmentally sustainable manner.

Public Communication

Public communication involves an overall strategy designed to establish mutually beneficial relationships with external organizations and individuals. Proactive public outreach regarding the benefits of IVM practices, along with program goals and maintenance objectives, including environmental stewardship, helps generate support for IVM programs and contributes to their success. The best public communication programs are mutually beneficial and should establish cooperative relationships with organizations such as nonprofits, community partners, environmental councils, natural resource agencies, youth groups, and community action committees that are dedicated to related causes. Participating in and supporting observances such as Arbor

Day, Earth Day, and Pollinator Week, community tree plantings, litter cleanups, and other ecologically centered activities can be beneficial.

Public communication is also a way to get students interested in vegetation management. Targeting young people can lead to the formation of lifelong allies who understand the purpose of IVM. Better still, some will undoubtedly become interested in pursuing IVM for their education and profession.

Internal Communication

Internal stakeholders are employees of the responsible organization, particularly those directly involved with vegetation management as well as contractors, executives, engineers, public relations departments, operations managers, call centers, and other staff members. Personnel should emphasize the importance and benefits of implementing IVM best practices. This is crucial because, without the appropriate information, people within an organization may not understand the complexity, public relations effects, and value added by the program. Furthermore, those not directly involved can help set priorities, anticipate and prevent potential problems, expand the communication network, and provide historical perspectives. Communicating during work can add a margin of safety; by knowing there is a project underway, operations staff may be able to respond more quickly to complaints and safety incidents than they would if they were unaware of the operation.

Importantly, communication within the organization needs to be clear and concise to ensure everyone understands the desired results. In addition to emphasizing the importance and benefits of IVM best practices, discussions should review process flow, particularly as it pertains to collaborating with other departments. Specifications and performance goals should delegate decision-making authority throughout the organization, as appropriate.

Quality Management

Quality management is a program that establishes how quality policies will be performed. It involves quality assurance (QA), quality control (QC), and independent assurance (IA). Quality assurance is a formal, management-level verification process that governs QC, acceptance, and IA and ensures the work meets program goals. Quality control is a maintenance-level system used by the contractor or field operations of the responsible organization to ensure their work meets project objectives. The intent of QC is to identify defects after work has been done but before a project is complete. It involves monitoring and sampling results, assessing the outcomes, and adjusting as necessary. It must be developed to be consistent with the QA process and with the resulting data verified by the management organization.

Independent assurance is a neutral, unbiased, independent evaluation of procedures used in accepting the final results. Its purpose is to establish the accuracy of the data used by those with responsibility for accepting work outcomes. Acceptance is a process used to evaluate how well results comply with specifications. Independent assurance is conducted by people outside of the program but may be either inside or outside the responsible organization. Those conducting IA must have the competencies to make determinations regarding the quality of work.

Vegetation Maintenance Plans

Vegetation maintenance plans apply to specific projects on a short-term time frame within the parameters of the vegetation management program. They involve setting objectives, defining action thresholds and tolerance levels, selecting control methods, and implementing QC and continuous improvement at project levels. Maintenance plans require a statement of work, which should cover the specific tasks and deadlines necessary to complete a plan. Plans should consider local conditions, attributes, and circumstances and be consistent with specifications.

Planning involves scheduling tasks, equipment maintenance, supervision, reporting, record keeping, and contingencies to address unexpected developments. Some components of maintenance plans, such as communication and QC, are presented in the management plan. They are covered again in this section as they pertain to their application at a project level. The steps of maintenance plans are described in the IVM process flow chart starting at the "Create maintenance plans and objectives" box (Figure 1).

Establish Maintenance Objectives

Maintenance objectives establish desired outcomes on a project level consistent with stated program goals. They should be clearly defined, documented by the vegetation manager, and based on the intended use of the site. The idea is to generate objectives that are precise and that explain exactly what needs to be done, who needs to do it, and where it needs to be accomplished. These elements are compiled in a statement of work.

Objectives should be SMART: specific, measurable, attainable, realistic, and timely. They need to be measurable so progress can be impartially determined. Objectives should be based on site factors, such as vegetation type, in addition to the available human, equipment, and financial resources. Objectives will vary from project to project, depending on logistical, topographical, environmental, fiscal, social, and political considerations. However, where it is appropriate, the overriding focus should be promoting environmentally sound, sustainable, site-specific plant communities that advance program goals. Ensuring timeliness requires setting deadlines to drive completion. Examples of project vegetation maintenance objectives can be found in Table 3.

Perform Assessments

Assessments are site and vegetation evaluations. They form the basis for the statement of work, including selection of control methods to be applied to a project. Assessments provide information at the time they are conducted. Components of assessments are presented in Table 4. Careful preparation is needed to ensure that valuable time and resources are directed toward obtaining useful information but are not wasted collecting unnecessary details. Because information can become quickly out of date, assessments need to be reviewed and updated whenever a project site is revisited. Quality standards should be established during planning to ensure precision in data acquisition.

Table 3. Examples of vegetation maintenance objectives.

- Complete project on time and within budget
- Conserve threatened and endangered species
- Control incompatible vegetation
- Enhance public and worker safety
- Ensure clear lines of sight on electric, pipeline, railway, and roadway ROWs
- Identify and remove high-risk trees
- Improve electric service reliability
- Maintain good relations with stakeholders
- Manage pollinator or wildlife habitat
- Reduce wildfire risk
- Optimize maintenance cost
- Protect sensitive sites
- Promote compatible vegetation
- Protect infrastructure integrity

Table 4. Examples of assessment components.

• Access routes • Archeological and cultural sites • Density and location of compatible and incompatible vegetation • Labor and equipment resource availability • Land ownership and use • Presence of species of concern • Riparian areas	• ROW width • Safety concerns • Stocking (density and height of compatible and incompatible vegetation) • Topography • Tree risk • Vulnerable or protected areas • Workload for various control methods

Site parameters can include elements such as location, property ownerships, topography, slope, key biodiversity sites, protected areas, soil health, land use, political considerations, human and financial resource availability, the presence of cultural and archeological sites, threatened and endangered species, and other attributes that could have a bearing on maintenance objectives. Vegetation assessments evaluate data on an array of vegetative characteristics, such as species abundance and compatibility, location, height, density, size, condition, anticipated growth rates, workload, and risk, considering action thresholds, tolerance levels, and the effectiveness of past control methods. Vegetation should be identified, mapped, and monitored for factors such as predominant species, anticipated growth rates, and habitat.

Remote Sensing

Remote sensing is obtaining data about subject parameters without making physical contact with the site. It is increasingly used to evaluate workloads. There are several types of remote sensing.

Cover-type mapping uses aerial photographs or satellite imagery augmented by ground checks to determine the nature of plant communities on a site. Internet sites are available that provide satellite photos, which can be applied to cover-type mapping when overlaid on geographic information system (GIS) platforms. An example is the Coastal Change Analysis Program from the United States National Oceanic Atmospheric Administration, which offers satellite imagery data for coastal regions of the United States on one- to five-year cycles. Ground checks (truthing) are necessary to confirm landform, vegetation, and soil attributes as well as aquatic or riparian delineation. The ground checks can also provide specific data to verify general information found on the maps (species, ecotypes, and understory, for example) and to verify the accuracy of older photographs.

Light detection and ranging (lidar) is another tool that is regularly used in vegetation management site assessment. It transmits laser pulses toward a target, records the time it takes to bounce back to sensors, and translates the results into a three-dimensional, virtual image (Figure 2). The technology can acquire data by air, ground, or both and can be combined with GIS to accurately locate vegetation on maps. It is precise and can document the exact location of vegetation and its relative distance from potential obstructions. It can also identify trees that are within striking distance of targets.

Sample Assessments

Sample assessments are vegetation evaluations that obtain data on a proportion of a population. They provide an estimate of attributes, such as workload and the condition of vegetation relative to action thresholds. If done properly, they deliver statistically valid information that can be used to develop maintenance plans. Generally, the larger the sample size, the greater the level of confidence in the assessment's accuracy.

Tree Risk Assessment

Many trees have defects that are impractical to detect, and it is not possible to predict when a tree will fail. However, by utilizing a systematic process of tree risk assessment and mitigation, threats to targets can be reduced.

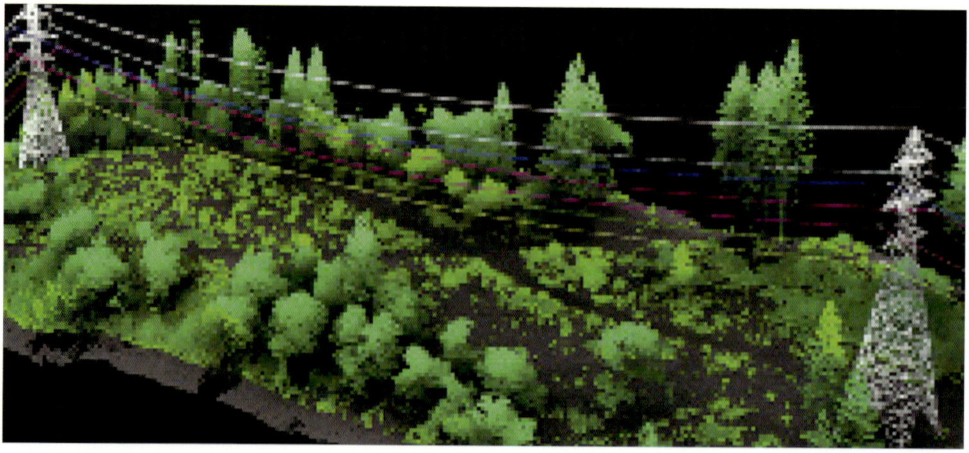

Figure 2. Lidar image.

Organizations should develop and implement plans to inspect their systems for trees that potentially pose unacceptable risk.

A Level 1, or limited visual, assessment is performed from a one-sided, specified perspective to identify obvious tree defects that could lead to failure and impact to targets. This type of inspection is recommended for large populations of trees typically found along many ROW corridors. Level 2, or basic, assessments include a 360-degree walk-around, are more detailed, and should be reserved for trees that are identified in the Level 1 inspection as requiring further scrutiny. Private property owners restricting access to trees outside of the ROW can hinder Level 2 assessments. Challenging topography and other obstacles may also impede access. Interested readers should refer to the American National Standards Institute standards for *Tree Risk Assessment* (ANSI A300, Part 9), the ISA's Best Management Practices: *Tree Risk Assessment*, and the ISA's and Utility Arborist Association's (UAA) Best Management Practices: *Utility Tree Risk Assessment*.

Determine Whether Action Thresholds Have Been Met

Assessment results should be used to determine whether vegetative conditions meet action thresholds. If they do, maintenance projects should proceed. If not, the project should remain in the planning stage until action thresholds are met.

Select IVM Control Methods

Control methods for IVM are maintenance procedures prescribed to achieve objectives (Table 5). The most appropriate methods are those that in the judgment of a vegetation manager are best suited to vegetative conditions, site-specific elements, land rights, applicable laws, regulations, and other factors identified in assessments. A diversity of methods should be evaluated for each site or project, and multiple control methods will often be integrated to achieve desired results. The ultimate outcome of IVM should be sustainable, compatible plant communities that exert biological control and minimize negative effects to ecosystems. Control methods will be discussed in more detail later.

Control methods should follow specific prescriptions for the vegetation and site characteristics on an individual project within program parameters. Large areas should be divided into manageable units, each clearly delineated according to observed site and vegetative conditions.

Table 5. Types of control methods.[1]

• Biological: management of vegetation by establishing and conserving compatible plant communities using competition, allelopathy, animals, insects, or pathogens
• Chemical: management of incompatible vegetation using herbicides or plant growth regulators
• Cultural: compatible land uses that hamper establishment and growth of incompatible vegetation—for example, agricultural systems such as crops and pastures, parks, or other managed landscapes
• Physical: management of incompatible vegetation through manual and mechanical techniques – Manual methods: management of vegetation using hand-operated tools such as handsaws and small power tools – Mechanical methods: management of vegetation using equipment-mounted saws, masticators for woody vegetation, mowers for herbaceous plants, or other devices
• Prescribed fire: a planned, controlled fire used to meet management objectives

[1] Control methods are not mutually exclusive. Often multiple control methods need to be integrated to achieve objectives. For example, chemical control often facilitates biological control.

Write a Statement of Work

A statement of work provides detailed expected outcomes on a project. It describes what will and will not be done. It should include project objectives, a timeline, a description of control methods and where they should be applied to the project, data to be collected, and required results. It should be written clearly, without jargon, and be consistent with program specifications and goals.

Communicate with Stakeholders

The communication strategy established in the vegetation management program is implemented in the maintenance plan. Landowners, tenants, and others who will be directly affected should be notified of upcoming work. Furthermore, members of the vegetation management team, including crew members, should know the facts regarding the project and understand its basic principles so they can respond to fundamental questions and know the proper channels to refer to for more complex issues. Communication should begin well in advance of work. It should be transparent, explaining the purpose of the program and the objectives of the project. Modifications may be implemented to address reasonable issues, provided they do not conflict with maintenance objectives.

Communication among vegetation managers and contract general forepersons, supervisors, and workers should be both written and verbal. Written instructions should include the information needed to successfully complete a project, including specifications; the statement of work; details about customers, property owners, or other stakeholders that have concerns; locations of environmentally sensitive or archeological areas; pertinent regulations; and any other considerations of consequence. Moreover, there should be plans for debriefings to review challenges and lessons learned for process improvement.

Communicating Outcomes

Both internal and external stakeholders should be kept apprised of vegetation management progress and outcomes. Communication should be honest and transparent, relating both successes and failures. Field visits during work are often appropriate to ensure managers understand how the project is proceeding and how results are developing so they can communicate program effectiveness throughout the organization and provide instructions to adjust to unanticipated outcomes.

Schedule and Perform Work

Control methods suitable for a project should be selected, scheduled, and implemented. The schedule should conform to the timeline established in the statement of work.

Perform Quality Assurance and Quality Control

At a project level, QC monitoring and testing should begin early to correct any possible miscommunication or misunderstanding on the part of crew members. Vegetation management work defects might be caused by misunderstandings due to inadequate communication; training deficiencies; weak supervision; inappropriate tools, equipment, or chemicals; or inadequate crew performance. Early and consistent observation also provides an opportunity to modify work practices, if need be, in time for a successful outcome. If QC identifies unacceptable variances from expectations, prompt corrective measures can be taken to bring them quickly back in line. Appropriate action should be directed at correcting any deficiencies, whether shoring up communication, strengthening supervision, improving training, supplying appropriate equipment, or making staffing changes.

Independent assurance (IA) follows QC and includes post-control reviews that require data collection on plant community changes, considering both compatible and incompatible height and densities, wildlife, water quality, stakeholder costs, and other factors.

Acceptance is among the final steps in the maintenance plan. It examines and verifies the results of QC and IA through sampling testing and inspections. Acceptance decisions are based on preestablished performance indicators. If variances from specifications exceed preestablished acceptance thresholds, work should be reassigned to the contractor or field operations to be corrected. If not, the work is accepted as complete.

Record Data

Accurate work records are necessary for adaptive management. Pertinent data (Table 6) can be recorded on web-based, GIS-based management software, which enables direct communication among crews, supervisors, and management. Herbicide records are required by law. Applicators should identify themselves and note the herbicide trade name, the active ingredient, and, in the United States, the EPA number. Applicators also need to track the amount of herbicide applied, the location of the control, weather conditions at the time of control, and how many trees or acres were treated, among other factors. Closed chain of custody best practices facilitate record keeping (see Appendix 7).

Table 6. Examples of information taken for records.	
• Date and nature of stakeholder communication	
• Identity of crew leader	
• Labor hours devoted to various activities – Chemical application – Cleanup (chipping, lopping, and scattering, removing slash from stakeholder property) – Inspecting	– Masticating – Mowing – Pruning – Removals – Stakeholder contact
• Location of work	
• Species of compatible vegetation retained that have been managed for	
• The number of crew members	
• The size of trees pruned (diameter at breast height)	
• The size of trees removed (diameter at breast height)	
• Unit area of compatible vegetation being promoted	
• Unit area and location of incompatible vegetation removed	
• Unit area of specific control methods utilized (including equipment used)	

Adaptive Management (Continuous Improvement)

Each maintenance project causes vegetation and site changes. Those changes often result in success in achieving plan objectives and progress toward program goals. Sometimes the changes can be less effective. Either way, results should be observed, and lessons learned should be documented. Outcomes should be compared to objectives and used to modify plans for the next round of maintenance as necessary. The important point is that each plan cycle builds on the previous application to progress toward program goals.

4. IVM Application

IVM Control Methods

Control methods for IVM are procedures used to achieve maintenance objectives and management goals. Control methods include biological, chemical (herbicide and tree growth regulators), cultural, prescribed fire, and physical (manual and mechanical) methods (see Table 5). They should be selected considering the results of site and vegetation assessments. The most appropriate method or methods are those best suited to meet objectives given vegetative and site factors, environmental issues, stakeholder concerns, and other matters identified during assessments. A diversity of methods should be evaluated for each site or project. Large areas should be divided into management units. Individual site prescriptions may be used in other areas with similar conditions; however, the use of a single control method across an entire program or project is not an integrated approach. Rather, variable environmental and site conditions often demand integration of multiple controls to achieve desired outcomes.

Wherever possible, the purpose of IVM should be to assimilate biological methods into projects, with the goal of directing change to sustainable, compatible plant communities that optimize the use of a site, reducing the need for further intervention while cost-effectively honoring environmental and social sustainability.

Biological Control Methods

Biological control is management of vegetation by creating, enhancing, and conserving compatible, stable, tree-resistant plant communities using competition, allelopathy, animals, insects, or pathogens (Figure 3). Wherever possible biological methods should be preferred, as they have been shown to maximize the achievement of IVM goals while minimizing costs and reducing overall environmental impacts.

Cover-type conversion to early successional plant communities is a prominent biological control where low-growing cover is desired. A central premise of the technique is the development of a stable, early successional plant community that resists invasion of incompatible plants. Early successional

Figure 3. Biological control methods. Biological control uses allelopathy, animals, insects, or pathogens to develop a stable, early successional plant community that resists invasion of incompatible plants. Once established, biological control largely excludes incompatibles.

plant communities are those that become established following a disturbance that opens a relatively large space. They are dominated by shrubs, forbs, and grasses. In time, predominately native, compatible, tree-resistant early successional plant communities will stabilize in an area and can be maintained by suppressing succession with minimal intervention.

Application

Conversion from incompatible plant communities to tree-resistant, compatible cover types is often achieved over time through progressively selective IVM controls. For example, if incompatible vegetation is dominant on a site, the first stage could be reclamation. Reclamation is vegetation management in areas that have not been actively maintained over extended periods. Reclamation can involve nonselectively clearing the management area using physical or other controls. Successive stages might employ cultural, prescribed fire, herbicide applications, or other techniques targeted over time against incompatible species with increasing selectivity. Once a compatible plant community dominates a site, biological control largely excludes incompatibles through competition, with minimal human intervention. Success can be determined by tracking the abundance of compatible species, so it is important to identify and monitor plant species composition over time.

Emphasis should be placed on providing opportunity to naturally release dormant seeds of native, compatible species wherever they exist. Native seeds are desirable because they have the provenance for environmental conditions at their site (such as local soil and moisture regimes), so they are better adapted than plants from elsewhere. Native plants can also provide habitat for indigenous animal species, including pollinators. Many native plant communities can also add a variety of color to enhance the aesthetics of a site.

A few plants ecologically compete by releasing chemicals that suppress other plants growing near them. Known as allelopathy, this characteristic can be advantageously leveraged by selecting for and promoting compatible, allelopathic species, which may push out targeted incompatible types.

The mechanism of succession is not limited to plant competition. Herbivores, especially small mammals, contribute as well. Early successional plant communities create escape and nesting cover while providing habitat for these creatures. In turn, the herbivores help perpetuate early successional communities by consuming tree seeds and seedlings. A synergy develops whereby herbivores maintain their habitat while sustaining the compatible plant community. The result minimizes the amount of intervention needed to achieve objectives. When managing for wildlife, practitioners must be mindful that creating a niche for one species will often be at the expense of another. For example, establishing habitat for grassland-obligate birds will displace shrub and forest avian species.

Insects and pathogens have also been used as biological control methods. For example, common St. John's wort (*Hypericum perforatum*) is a perennial herb native to Europe, Africa, and the Middle East that is dangerous to livestock and invasive in North America. *Chrysolina* spp. is a genus of leaf beetles introduced to North America that feed selectively on St. John's wort and have proven successful in controlling this plant.

Advantages and Disadvantages

A central advantage of biological methods is that they use nature as an ally. Stable, compatible communities reduce the amount of work, including herbicide application, with each successive treatment. Once established, the early successional, compatible community can often be maintained indefinitely with minimal intervention and cost. The result is that biological controls simultaneously contribute a wide range of ecological benefits, social and economic value, and cost effectiveness because the resulting compatible plant community naturally does the work that would otherwise have to be achieved through intervention.

It should not be surprising that cover-type conversion delivers environmental benefits, given that it was developed by ecologists such as Frank Egler, William Niering, William Bramble, and William Byrnes. It has been demonstrated to enhance the richness of native plant species and to develop habitat for early successional wildlife such as songbirds, herptiles, pollinators, and other creatures. Recent work has focused on creating pollinator habitat and wildlife escape and nesting cover along roadside, electric, pipeline, and other ROWs. In many cases, habitat for these animals and beneficial insects has been dwindling due to development, agriculture, and other practices, some of which favor trees. So, rather than being sacrificed because it cannot accommodate trees, land subject to cover-type conversion provides areas of opportunity to establish and conserve beneficial habitat that might be otherwise limited.

If misapplied, biological control can have a negative impact on the environment. For example, some aggressively invasive shrubs like multiflora rose (*Rosa multiflora*), forbs such as garlic mustard (*Alliaria petiolata*), and grasses such as reed canary grass (*Phalaris arundinacea*) are low growing and therefore can be construed to be compatible with some narrow maintenance objectives, like keeping vegetation from interfering with electric facilities. However, managing for such environmentally ruinous species is not an IVM best management practice, regardless of its efficacy for achieving other outcomes. For comprehensive lists of invasive weeds, consult local plant authorities.

A disadvantage of insect agents is that they have not been identified for many incompatible species. Further, before non-native insects are released into the environment, expensive, time-consuming research must be done to make sure they will not cause unintended environmental harm. This can happen when indigenous wildlife species become reliant on the invasive plants after they have displaced the native plant community. For example, willow flycatchers (*Empidonax traillii*), which are endangered in some regions, have become dependent on invasive saltcedar (*Tamarix* spp.) for nesting habitat in the southwestern United States. *Diorhabda*, a central Asian genus of leaf beetles, have proven successful in controlling saltcedar; however, in controlling them, *Diorhabda* also limited the nesting sites of willow flycatchers, thus causing unintended harm. Finally, insects cannot be contained on-site, so they may migrate to off-target areas, which might have negative environmental consequences.

Another disadvantage of biological methods is that their mode of action (MOA) is relatively slow compared to other controls. Consider that it can take many years to completely convert an area dominated by incompatible plants into a stable, compatible plant community. Stakeholders must be informed that other control methods will be employed while biological controls develop on the site. It is important to communicate the long-term benefits to stakeholders so they understand the time involved.

Chemical Control Methods

Chemical control methods involve management of vegetation using herbicides or tree growth regulators (TGRs). Herbicides are chemicals that control plants or plant parts by interfering with specific botanical biochemical pathways. Tree growth regulators are substances designed to reduce vegetative growth rates by interfering with natural plant processes; they are actually plant growth retardants, though they are not marketed as such. Preference should be given to chemicals that minimize risk to humans and the environment.

Herbicides and TGRs must be applied by qualified applicators according to label directives. Applicators are required to read and comply with label instructions as well as all other laws and regulations pertaining to chemical use. Label instructions for personal protective equipment are also important. More information on chemical use can be found in Appendix 7.

Application

Wherever possible, the goal of chemical control methods should be to minimize their use over time by facilitating cover-type conversion. That means promoting compatible plant communities by selectively applying chemicals to incompatible plants that are prone to resprout or sucker after removal. When trees that resprout or sucker are removed without herbicide treatment, dense thickets can develop, impeding access, swelling workloads, increasing costs, blocking lines of sight, and deteriorating early successional wildlife habitat (Figure 4). Treating incompatible plants with herbicide creates an opening for early successional or other compatible species to develop and outcompete incompatible species, ultimately reducing workload. Consequently, using herbicides to achieve cover-type conversion has been described as chemically facilitated biological control.

Selectivity is a function of herbicide type and application technique. There are a variety of herbicides, each of which have different MOAs depending on the formulation and characteristics of the active ingredient. Selective herbicides only control specific kinds of plants when properly applied. For example, synthetic auxins are a class of selective herbicides that kill broadleaved plants but do not harm grass species when used according to the label. By contrast, nonselective herbicides work against both broadleaved plants and grasses. Selective applications kill specific plants or pockets of plants.

Nonselective techniques target areas rather than individual plants. Nonselective use of nonselective herbicides controls all treated plants. Nonselective

Figure 4. Sprouting from untreated physical control. When trees that resprout or sucker are removed without herbicide treatment, dense thickets can develop, impeding access, swelling workloads, increasing costs, blocking lines of sight, and deteriorating early successional wildlife habitat.

techniques are only appropriate where the objective is removing all the plants in the management area, such as in and around electrical substations or the ballast zone of railroad ROWs. Nonselective use of a selective herbicide controls treated plants that are sensitive to the herbicide (most commonly broadleaved species) without affecting other plants. Selective use of either herbicide affects individual plants that have been treated. Selective use is preferable unless incompatible-vegetation density is high.

For management areas overgrown with incompatible vegetation, initial prescriptions often combine chemical with physical or other methods. In some cases, nonselective application of selective herbicides is the most appropriate initial IVM methodology. Subsequent treatments should be increasingly selective. Selective herbicide applications can target seedlings of non-native, invasive, or other incompatible plant species that might germinate from the reservoir of seeds in the management area. It could take repeated applications over several years to deplete the incompatible seed bank. See Appendix 7 for more information on selectivity.

Tree growth regulators can be helpful by slowing growth rates of some trees or other woody plants where removal or cover-type conversion is prohibited or impractical. A common example is fast-growing tree species located in proximity to urban electric distribution lines. TGRs are also a tool to consider in difficult-to-access areas or where there are significant safety issues. They have been used on highways, in wildlife habitat, and in other areas for suppressing seed heads on grasses and minimizing seed production and distribution of undesirable species.

Advantages and Disadvantages

An advantage of herbicides is that, when properly applied, they are effective and efficient. Herbicides kill roots, so the entire plant is controlled, eliminating resprouting. They can be essential for reclamation; they are also helpful in cover-type conversion or prairie restoration by mimicking the effects of fire. By promoting cover-type conversion, herbicides minimize soil disturbance that would be caused by methods that rely on heavy equipment, enhance early successional wildlife diversity by improving forage, and promote escape and nesting cover for herbivores and pollinators. Herbicide use creates openings that release plant species of concern, including threatened and endangered species, from seeds that lay dormant in the soil. They can also be effective in controlling noxious weeds. From an economic perspective, herbicide use in conjunction with physical methods to facilitate cover-type conversion has been demonstrated to reduce costs 50 percent over time compared to physical methods alone, with lower carbon emissions, less disturbance to the soil, and reduced disruption to wildlife habitat.

Herbicides and TGRs must be applied correctly and in the proper doses to be effective. It is necessary to inspect areas after they have been treated to determine whether the targeted vegetation was controlled. Moreover, vegetation treated in leaf will brown out, which may cause public concern. Misused herbicides cause environmental risks due to drift, leaching, volatilization, and persistence within a habitat, which can impact nontargeted vegetation, wildlife, and water resources. For example, repeated broadcast spraying can alter soil chemistry and harm compatible species, which impedes cover-type conversion. Also, some herbicides can persist in water, having adverse effects on aquatic fauna such as fish and amphibians.

Herbicide use can also be controversial. For example, many people may be philosophically opposed to chemical use. Their perspective must be considered, which means opening dialogue to understand concerns (see Communication and Stakeholder Engagement section in Chapter 3).

Cultural Control Methods

Cultural methods manage vegetation by establishing and maintaining compatible land uses that preclude the growth of incompatible vegetation. Examples include agricultural systems such as crops and pastures, targeted grazing, and parks or other managed areas.

Application

Agricultural systems are common in rural areas and can be leveraged as sustainable compatible plant communities. In addition to agricultural systems, cultural methods such as cultivating and hydroseeding can be used to initiate cover-type conversion where dense, incompatible vegetation dominates a site and needs to be removed and replaced. Prairie establishment is an example that requires considerable site preparation before planting. Aftercare can involve an extended integrated strategy of prescribed fire, mowing, targeted grazing, or selective herbicide application. Once the prairie becomes established, it naturally resists incompatible plants and develops into a biological control system.

Cultural control can also accommodate multiuse repurposing. For example, parks can simultaneously facilitate the intended use of the site (such as space for electric or pipeline corridors) and provide social benefits attendant to the green space. Rails-to-Trails Conservancy is an example of repurposing. This nonprofit converts abandoned railroad ROWs to trails. In sections accessible and open to the public, some of these ROWs allow for targeted community outreach with interpretive signs and other techniques. For instance, signs can introduce interested parties to the benefits of pollinator gardens, wildlife habitat enhancement, and other environmental improvements while meeting primary program goals. However, vegetation managers need to be mindful that in many cases the right-of-use does not permit public access. Practitioners may not manage for trails, interpretive areas, or other social applications without authority to do so.

Seeding

Seeding is an important cultural method. Native seed mixes carry advantages. For example, they contain a diversity of species that are adapted to a variety of local growing conditions and ecological niches, including seeds of plants suited to different moisture, light, soil, temperature, and other conditions. Seasonality of blooming should be a selection consideration to maximize aesthetics and availability to pollinators. Selecting seeds requires knowledge of the growth requirements of the compatible plants being planted. Cover crops of quickly germinating, often annual species can be desirable to temporarily hold a site until permanent cover can become established. Cover crops planted along with perennial species are called nurse or companion crops and are used to stabilize an area (e.g., by preventing erosion) until the permanent species become established. Information on commercial seed labels can be found in Table 7.

Table 7. Information on commercial seed labels.

- Common and technical names of the seeds
- Name and number of noxious weed seeds per unit weight
- Percentage of the material that is the intended seed
- Percentage of material that is plant debris or other non-seed matter
- Percentage of seeds that are non-weed species
- Percentage of weed seed
- Percentage of seeds that will sprout readily in a germination chamber
- Percentage of seeds that will not germinate readily due to having a hard seed coat, requiring weathering in the soil, or needing to be pretreated
- Pounds (kilos) of live seed (weight of live seed × % purity × (% germination + % dormant))

Seed beds often need to be cultivated and raked in preparation for planting. There are three methods of seeding: drill seeding, broadcasting, and hydroseeding. Drill seeding uses a machine to insert seeds into the ground at a prescribed rate and depth. It works best on flat areas or gentle slopes; multiple passes often provide the best results. Drill seeding is generally preferred for native grasses. Broadcasting can be done with a mechanical spreader for large areas or by hand for more compact projects. For best results, the seed should be mixed with sand or a similar carrier. Hand seeding should be followed by light raking. Rolling or otherwise lightly packing broadcast seed may also be beneficial. Hydroseeding is a type of broadcast seeding that uses a seed-mulch slurry mix. The mix is broadcasted onto a target area (Figure 5). It is effective on steep slopes where conventional broadcast seeding is impractical. Hydroseeding reduces erosion and the mulch can be colored so applicators can easily determine their coverage.

Targeted Grazing

Targeted grazing is using domestic animals to feed on incompatible vegetation during vulnerable stages in its life cycle. While the use of livestock might seem to be a biological control, it is considered cultural because the animals are brought in and actively managed. On the other hand, a stable wildlife population is a biological control because it doesn't require such direct human

Figure 5. Hydroseeding.

intervention. Targeted grazing is more sophisticated than simply releasing livestock into a management area. It requires an understanding of the animals as well as the characteristics of the compatible vegetation being managed for and the plant species being controlled. Proper timing and intensity put desirable cover types at an advantage over vegetation targeted for control. For example, many plant species are compromised if they are defoliated during flowering and seed set, when their energy resources are directed at reproduction. Defoliation during that time weakens them and leaves them vulnerable to competition or disease. Targeted grazing is thus often best implemented when incompatible species are in flower and when compatible species are not.

Some departments of transportation have deployed goats on road ROWs. Goats often not only preferentially eat weeds but also sterilize some weed seeds that pass through their digestive systems, effectively inhibiting the ability of the weeds to dominate a site. Goat grazing for invasive plant management also boosts organic matter, decreases erosion, and increases desired plant species diversity. Targeted grazing is most effective when integrated with other controls, such as physical or chemical methods.

Advantages and Disadvantages

Cultural control can effectively provide stable, compatible plant communities. Agricultural systems or parks with turf and low-growing ornamentals are dependable control methods as long as the land is devoted to crops or green space. Moreover, seasonal (i.e., phenological) grazing, masticating, and mowing can eventually result in cover-type conversion with its attendant environmental advantages such as promoting pollinator habitat and early successional wildlife habitat.

A disadvantage of cultural methods is that without vigilance, incompatible species can be introduced into spaces that had once been managed for alternative uses such as parks or agricultural fields. For instance, there have been cases where, due to a lack of understanding among municipal public works employees, trees have been inappropriately planted over pipelines or under power lines, which compromises the critical infrastructure that is the primary use of the site. Further, land uses may be converted over time and become incompatible with the site. So, practitioners should not assume that a given compatible use will exist in perpetuity.

A disadvantage of targeted grazing is that grazing animals do not always prefer the targeted vegetation. For example, in the Pacific Northwest of the United States, sheep do not eat conifers or red alder (*Alnus rubra*), making them incompatible with management objectives for electric, highway, and pipeline ROWs. On the other hand, in some cases, animals have been found to be nonselective and might consume compatible and incompatible species alike, making it difficult to achieve desired plant community conversion. Grazing typically does not kill plants, which may frequently resprout if not treated with herbicide. Other problems associated with managing grazing animals include handling fencing and herding, protecting them from predators, providing water and supplemental feeding, and preventing exposure to disease. Overgrazing can also potentially cause erosion, which can negatively affect water quality and soil health and conflict with big game, nongame wildlife, and pollinator habitat.

Physical Control Methods

Physical control is management of incompatible vegetation through the use of manual and mechanical controls. Physical controls can promote and enhance compatible vegetation, particularly when combined with other methods.

Manual Control Methods

Manual methods control vegetation using hand-operated tools, such as handsaws, mattocks, machetes, shovels, hoes, and rakes, and small power tools such as chain saws and string trimmers. Manual techniques include pruning, vegetation removal, girdling, pulling, and grubbing, among others.

Application

Manual methods can be used where others may not be practical, such as in cases with urban or developed locations, steep slopes, difficult-to-access places, or wetlands and other environmentally sensitive areas. Manual methods may also be applicable in small management areas where it is impractical or not cost effective to deploy equipment or apply other methods.

Pruning is a common manual technique. When pruning is necessary, it should be conducted according to the most current version of the American National Standards Institute *Pruning* standards (ANSI A300, Part 1) and the ISA's Best Management Practices: *Utility Pruning of Trees*. While aerial lifts are machines that provide tree workers access to trees, they are considered a manual control in this BMP because work from them is performed using handheld saws or other devices.

Girdling trees is a manual technique performed by cutting a deep ring around the trunk with a handheld saw or a chain saw. The goal of girdling is to kill the tree by severing the phloem and outer rings of sapwood to disrupt water and essential element and carbohydrate transport. While girdling kills trees, it leaves them standing for wildlife habitat. Other structurally unsound or dead trees can also be left for wildlife if they will not strike targets of consequence when they inevitably fall.

Advantages and Disadvantages

An advantage of manual techniques is that they are selective and can precisely target incompatible vegetation for control, allowing compatible vegetation to develop. Moreover, manual pruning is the only way to honor natural targets—branch collars and appropriate laterals—which limits the negative impact of pruning wounds. Further, manual methods allow for more targeted access and application where other control methods, such as physical, chemical, or prescribed burn, would be impractical.

The disadvantages of manual controls are that they can be slow and more costly than alternatives. They are only a temporary solution for removing woody species that are prone to resprout and form thickets, which eventually increases the volume of work unless combined with chemical or other controls. Manual techniques (such as pulling, hoeing, grubbing, and raking) are ineffective for large-scale noxious or invasive weed control. Further, manual control can be difficult to carry out in expansive, densely vegetated areas. In many cases, pruning is inappropriate where trees conflict with maintenance objectives. Gas-operated chain saws used for manual control can carry the disadvantages of possibly sparking during dry, hot periods and creating noise that disturbs people and wildlife. Electric models offer a practical alternative in such cases.

Mechanical Control Methods

Mechanical control uses equipment-mounted devices such as mounted saws, masticators, and mowers to manage vegetation. Masticators grind woody plants, while mowers cut herbaceous species.

Application

Mechanical controls are efficient and cost effective, particularly for clearing large, densely vegetated areas such as those frequently encountered in reclaiming extensive neglected or overgrown areas. Seasonal mastication and mowing means mechanically managing vegetation at specific stages in plant life cycles and requires an understanding of the species involved. The idea is to promote compatible species by timing masticating or mowing to maximize vulnerability of incompatible species and minimize it for compatible types. For example, proper timing might promote flowering and seed set in compatible species and discourage it in incompatible plants, giving a reproductive advantage to the compatible species. Mastication can also be done to reduce combustible fuel.

There are a variety of machines that can be used for IVM. Examples include:

- Masticators (Figure 6)—machines that remove and grind small woody plants and fell small trees.
- Mowers—machines that cut herbaceous plants. They can be used to reduce the height of herbaceous vegetation and for seasonal mowing.
- Shears—devices for whole tree removal mounted on heavy equipment. Shears can fell, lift, and stack trees.
- Mechanical pruning machines—all-terrain vehicles equipped with a boom that extends a saw (Figure 7). Mechanical pruning can also be done with an array of blades slung beneath a helicopter.

Advantages and Disadvantages

The most important advantage of mechanical control is that it is fast and less expensive than manual control alone, so it is suitable for large areas. For example, mechanical pruning devices can do the work of as many as half a dozen tree crews. Masticating and scattering can improve aesthetics, facilitate debris decomposition, reduce fuel loads, and minimize fire risk. Appropriate timing and frequency can affect plant community development, so seasonal mowing and mastication combined with chemical control can contribute to cover-type conversion. To safely achieve desired outcomes, machinery must be run by skilled operators.

A disadvantage of mastication is that it is nonselective, and by itself, it seldom converts incompatible plant communities to compatible cover types. The

Figure 6. Masticator mounted on an excavator.

problem is that masticated vegetation still retains its roots from which shoot growth can proliferate, thus increasing stem densities in many species (Figure 4). So repeated cutting is only effective when combined with herbicide application. Mastication and mowing can also spread noxious and invasive weeds, particularly if improperly timed.

Mechanical controls can create disturbances, potentially causing soil erosion and damaging sensitive areas such as wetlands and cultural sites. Masticating to too short of a height can damage desirable plant species, and harm soil, including leaving bare soil patches, that can lead to erosion and potentially create a seedbed for incompatible plants. There may also be restrictions on use of heavy equipment in certain areas or at certain times. For example, there may be seasonal restrictions to prevent harm to nesting wildlife. Machines can generate noise that can distract or frighten animals. At times, machines can leave petroleum products behind from normal operation or from leaks and spills, particularly if they have been poorly maintained. Furthermore, heavy equipment can be risky on steep terrain, where it can be unstable.

Figure 7. Mechanical pruning machine.

Mechanical cutting also carries inherent safety concerns due to sharp cutting blades and rapid discharge of severed wood. Additionally, it is difficult to make good pruning cuts with mechanical pruning machines; the resulting wounds can be damaging, so machines are inappropriate for high-value or landscape trees. It is important to have mitigation or abatement plans as contingencies to offset these disadvantages.

Prescribed Fire

Prescribed fire is management of vegetation using a planned, controlled fire. When done correctly, prescribed fire can facilitate cover-type conversion by consuming woody and invasive plant species while promoting grasses and forbs (Figure 8).

Figure 8. Prescribed fire.

Application

Low-intensity fire is part of the ecology of many habitats, including prairies, meadows, savannas, and some forest types. Those plant communities naturally benefit from prescribed fire. Prescribed fire is most often effective when combined with physical or chemical control methods. This is particularly the case for areas that have been neglected or have undergone years of mismanagement. Prescribed fire is not a one-time control method but a continual practice. Fire-frequency schedules must be driven by program goals in addition to site and vegetative factors. However, a return rate of at least once every three years is often necessary to control woody plants.

Detailed planning is necessary for prescribed fire. Vegetation density and height, moisture content, topography, the proximity to development, weather conditions (including wind speed and direction, humidity, and temperature), time of year, contingencies for containment, and other factors need to be considered—not only for effective control but also to ensure fire does not spread into off-target areas.

Advantages and Disadvantages

Prescribed fire effectively restores the historical dynamics of fire-derived ecosystems and cost-effectively keeps plant communities in early successional stages. It promotes cover-type conversion and the advantages that brings. Fire can promote earlier green-up of herbaceous plants in the spring because charred surface organic matter absorbs heat from the sun and warms the soil. Moreover, prescribed fire quickly recycles essential elements and can help control disease and insect pests, including parasites like ticks. Prescribed fire can also moderate high-intensity fire risk by consuming dry combustible vegetation, thereby reducing fuel levels that might otherwise generate potentially catastrophic fire. In that regard, prescribed fire has been applied in the urban-wildland interface to protect houses and other infrastructure. Low-intensity fire that consumes the mulch layer increases the amount of bare ground and can boost seed availability to upland birds and small mammals. Invasive or noxious plant species that degrade natural habitats or otherwise interfere with management goals can be removed by seasonal burning, where fire is timed just before seed set.

From a negative perspective, if a target area has been neglected or mismanaged, prescribed fire is impractical, as it could result in a high-intensity burn that is difficult to contain. Further, if the interval between applications is too long, many woody species will grow too large to be controlled by low-intensity fire. Prescribed fire can be difficult to manage in narrow, linear management areas like roadside, electric distribution line, or railroad ROWs. Fire is an inappropriate tool on petroleum pipelines.

5. Economic Viability

Economic viability means that there must be a long-term commitment to cost-effective implementation of IVM, which includes stable, adequate funding. Along with that comes a responsibility on the part of practitioners to deploy cost-effective policies and procedures. That often means herbicides should not be relegated to a last resort but rather accepted as an integral part of the IVM toolbox. For example, mechanical- and herbicide-based prescriptions are significantly less costly than repeated mechanical techniques—nearly 50 percent less annually once cover-type conversion has been established. The point is that implementing the highest value control for the site often calls for multiple methods. It also requires implementing the systematic IVM process, featuring adaptive management.

6. Environmental Stewardship

Environmental stewardship is a central tenet of IVM. Best practices of IVM consider the direct and indirect environmental impacts of the program, both positive and negative. Wherever possible, IVM should be designed to promote early successional wildlife such as grassland birds, many pollinators, small mammals, and other species. Managers should be aware of species requiring priority for conservation measures—including rare, threatened, endangered, or otherwise protected species—and should avoid harming them. If habitat for these animals must be disturbed, practitioners need to comply with applicable laws, guidelines, and regulations. For example, in the United States, vegetation management on federal lands is subject to NEPA. NEPA is applicable where approval of specific projects, such as construction or management activities located in a defined geographic area, is necessary. The process can involve environmental impact statements or environmental assessments.

An important initiative for pollinators from the Rights-of-Way as Habitat Working Group (RHWG) promotes Candidate Conservation Agreements with Assurances. These assurances are voluntary conservation agreements between the US Fish and Wildlife Service and other entities, including state and local jurisdictions, in addition to private landowners. They are designed to address the needs of at-risk species before they become listed as endangered or threatened. The program is tailored to nonfederal landowners and managers who voluntarily agree to conservation measures intended to stabilize or restore the target species. In turn, the US Fish and Wildlife Service provides participating property owners and land managers with an Enhancement of Survival permit containing assurances that they will not be required to implement additional conservation measures beyond those in the agreement, even if the species is listed. By proactively promoting conservation action ahead of regulation, these agreements benefit both the parties entering into them and the at-risk species.

Wetlands and Streams

Wetlands should be worked using suitable control methods. Cover-type conversion can be applied in wetland or riparian areas to enhance wildlife habitat. If herbicides are to be applied, only those labeled for use over, near, or at water's edge may be used. In addition, vegetation maintenance may provide an opportunity to place cut or removed vegetation, including logs, into wetland areas and streams to create basking habitats for reptiles and eddies for fish, but only if authorized by a competent authority. To protect streams, incompatible vegetation may need to be selectively pruned, removed, or treated with an appropriate herbicide to gradually establish a compatible riparian plant community. Machinery may only use existing or designated stream crossings.

Wetlands, riparian areas, and surface water supplies, including reservoirs, drinking water wells, and springs, need to be protected by buffers. Buffer zones should retain as much compatible vegetation as possible, which often means the areas are worked with manual techniques and herbicides labeled for such areas. Machine work should be avoided in buffer zones, as equipment may disturb fragile soils, leading to erosion and sedimentation problems. Machines may also leak or spill petroleum products, causing pollution. Practitioners working along with competent authorities should determine the appropriate size for buffer zones and collaborate on selecting control methods.

When it is necessary to move heavy equipment into wetlands, efforts must be made to protect these delicate environments. In general, tracked machines distribute weight more evenly and are preferred over those with tires. When necessary, mats made from timbers or grids can be used to access sites inside of wetland areas. Disturbances to wetland areas must be minimized during installation and removal of mats. Equipment and supplies should not be stored within wetland areas.

Debris Disposal

Logs and cut vegetation that result from IVM operations should be handled in a manner consistent with applicable laws and regulations, adjoining land use, terrain, aesthetics, wildlife habitat, and fire risk. Logs may be recoverable for firewood or timber products and are often best left for the property owner or as wildlife habitat. Logs should not be moved from the work site if they are likely to be infested with disease or insect pests that could spread. Where appropriate (such as in remote areas or in wildlife management areas), dead standing timber that cannot strike valuable targets can be left as wildlife habitat, including for use by wildlife. Cut woody material can either be placed into piles, windrowed, or lopped and scattered. Some jurisdictions may limit the height and length of resulting piles. Cut vegetation or logs should not be placed in streams or where floodwaters can wash materials downstream, unless requested or authorized by a competent authority.

7. Social Sustainability

Social sustainability is a system of establishing healthy and livable communities for current and future generations. It focuses on understanding the effects operations have on people and society.

Environmental, Social, and Governance

Environmental, social, and governance (ESG) are essential criteria that are used to rate corporations and attract principled investors, which is becoming increasingly important in corporate governance. Integrated vegetation management programs contribute to the environmental component of ESG and can be reported on ESG indexes. Social initiatives involve diversity, human rights, and consumer protection. Corporate governance consists of management structure, worker protection, employee relations, and executive and employee compensation, among other matters. IVM managers should use the opportunities offered by ESG initiatives to communicate with stakeholders and advance the interests of both their programs and their parent organizations. Making IVM programs a critical element of corporate ESG initiatives requires coordination across departments, which can help stabilize budgets and ensure that IVM programs are recognized for the value they bring to their organizations.

An IVM program that uses best management practices provides organizations various opportunities to demonstrate their commitment to ESG criteria, including pollution prevention, natural resource management, habitat stewardship, biodiversity, endangered species protection, and other factors. Managers may use conservation measures to demonstrate goodwill and improve relations with regulatory agencies. For example, utilities have promoted wildlife habitat in one area to facilitate transmission line construction in others. IVM programs are also an opportunity to inform the public about environmental matters, connect with at-risk groups, or connect neighborhoods through ROWs. Right-of-Way Stewardship Accreditation can be used as evidence of a commitment to ESG criteria.

Archeological or Cultural Sites

Archeological or cultural sites demand respect as part of human heritage. Vegetation management activities should not disturb known archeological or cultural sites. In areas where cultural sites are known to exist, archeological resources should be located and marked, and a plan established to adequately protect them during work. Responsible organizations should cooperate with representatives of Indigenous groups to respect their rights and customs. Field data inventories of known sites should be kept on file. Practices that will not damage the sites, such as manual cutting and backpack herbicide applications, should be considered for use at these locations.

8. Summary

Integrated vegetation management is a system of managing plant communities in which compatible and incompatible vegetation are identified, action thresholds are determined, tolerance levels are established, and control methods are evaluated, selected, and applied to achieve management goals and maintenance objectives.

The IVM process is an iterative, strategic process applied tactically. It is comprised of both a vegetation management program, which is strategic and long term, and one or more vegetation maintenance plans, which are tactical and short term at the project level. The IVM process coordinates the vegetation management program and governs the vegetation plan in a documented procedure designed in a continuous loop. It requires an understanding of ecosystem dynamics and plant species that are compatible and incompatible with desired outcomes. With that in mind, the IVM process is cyclical because successfully managing dynamic natural systems must be ongoing. Known as adaptive management, the systematic, cyclical structure provides flexibility to adjust plans as circumstances evolve and as new information about changes introduced into the management area becomes available.

Communication and stakeholder engagement is an overall strategy for public relations and education. It is used to inform external and internal stakeholders regarding IVM. Communication is essential to planning and implementing a successful program. Proactive organizations anticipate stakeholders' interests and provide relevant information in a variety of formats. Honest, transparent information will build trust and serve in the long-term best interest of the operation.

Tolerance levels and action thresholds are central concepts to IVM. Tolerance levels are the maximum incompatible plant pressure allowable without unacceptable consequences. Plant pressures are factors such as vegetation height, location, density, level of incompatible species, condition, or a blend of factors. Unacceptable consequences may include safety incidents, compliance violations, economic harm, environmental degradation, or a combination of these or other adverse outcomes. Tolerance levels should never be exceeded. Action thresholds are levels of incompatible plant pressure where control measures should be implemented to prevent tolerance levels from being breached.

IVM control methods are maintenance procedures prescribed to achieve objectives. They include biological, chemical, cultural, prescribed fire, and physical (manual and mechanical) methods. The most appropriate controls are those that in the judgment of a vegetation manager are best suited to vegetative conditions, site-specific elements, land rights, applicable laws, regulations, and other factors identified in assessments. A diversity of methods should be evaluated for each site or project, and multiple control methods will often be integrated to achieve desired outcomes. The goal of IVM should be to promote sustainable, compatible plant communities that exert biological control and minimize negative effects to ecosystems.

Control methods should be selected based on site and vegetative conditions determined through assessments. Assessments are site and vegetation evaluations. They provide the basis for the statement of work, including selection of control methods to be applied to a project.

IVM best management practices include economic viability, environmental stewardship, and social sustainability. Economic viability means that there must be a long-term commitment to cost-effective IVM implementation, which includes stable, adequate funding. Management and maintenance plans must be adequately funded to achieve their goals and objectives. IVM best practices consider the direct and indirect environmental impact of the program, both positive and negative. Social sustainability is dependent on inclusion and on making stakeholder engagement an integral part of an IVM program.

Appendix 1
Electric Utility IVM

Unmanaged vegetation growing near utility ROWs can damage electric facilities and cause problems with safety, reliability, access, emergency service restoration, regulatory compliance, security, inspection, and other factors. It can also compromise conformity with environmental, legal, regulatory, economic, and other imperatives. Electrocution due to contact with high-voltage lines is among the most common causes of arboricultural workplace fatalities. Further, every year, members of the public are injured, maimed, or killed due to electric shock suffered from contact with high-voltage lines they accessed through trees.

Vegetation interference with power lines causes between 20 and 30 percent of electric outages on distribution systems in North America and has been responsible for initiating ruinous transmission grid failures that have subjected millions of people to prolonged power loss. The most notorious example is the August 14, 2003 catastrophic cascading blackout in eastern North America that left 50 million people without electricity, some for weeks. Vegetation can cause electric service interruptions by mechanically damaging facilities or creating electrical faults. Mechanical damage can happen when trees or tree parts fall into or through conductors. Trees contacting power lines can produce faults by weakening conductors or causing a high-voltage flashover. Moreover, during dry conditions, sparks caused by vegetation contacting power lines can start wildfires. Keeping vegetation clear of power lines can help mitigate these and other risks.

Some governmental regulations require minimum clearances between vegetation and high-voltage lines. For example, the North American Electric Reliability Corporation (NERC) Transmission Vegetation Management Program standard requires minimum clearances for critical transmission lines. Moreover, in the United States, the Energy Policy Act of 2005 contains provisions for electric system reliability standards, including those for vegetation management. Based on this provision, the Federal Energy Regulatory Commission (FERC) has adopted the NERC Transmission Vegetation Management Program standard, which essentially gives the NERC standard the force of law. To reduce the risk of wildfire and to improve reliability and safety performance, some states require utilities to maintain minimum clearances for distribution voltages. Another important American standard is the National Electrical Safety Code, section 218 of which requires utilities to prune or remove trees that may damage ungrounded supply conductors.

Clearance Requirements

Minimum vegetation clearance distances (MVCDs) from electric facilities may be established by individual utilities or regulatory oversight. They serve as tolerance levels for utility vegetation management. When creating MVCDs for energized conductors, practitioners should consider the following elements:

- The potential growth of vegetation in the area
- The combined movement of vegetation and conductors in high wind
- The sag of conductors due to elevated temperatures or icing
- Regulatory clearance requirements

Utility vegetation managers should be aware that transmission lines are subject to sag and sway, particularly midspan. Lines sag when they are heated due to heavy electrical loads, high ambient temperature, or both (Figure 9). Conductors sway under windy conditions, more so where the lines have sagged (although wind can cool conductors, reducing sag to some extent). Practitioners should establish MVCDs with the understanding that lines can be positioned at any point along their engineered positional continuum, including maximum sag and sway. Action thresholds for MVCDs need to be established from the extremities of that continuum.

Vegetation managers must be mindful that IVM requires a more proactive approach than simply maintaining minimum clearances. IVM best practices are designed to prevent the establishment of incompatible vegetation while helping compatible vegetation to thrive. Trees that have grown to the point where they encroach on tolerance levels or could cause an outage at any moment indicate a breakdown of the IVM program. Action thresholds should be established to initiate work long before vegetation has the potential to violate tolerance levels. Achieving only mandated MVCDs (such as those in FAC-003), while technically in compliance with regulations, is not a best practice, nor should minimum clearances be used as a limitation for managing vegetation on a ROW or for evaluating the efficacy of IVM operations. Doing so allows establishment of incompatible species in the ROW, which generates the need for periodic topping or severe pruning. This not only fails to effectively prevent risk to the lines but also potentially damages trees. Moreover, tree maintenance under such conditions can unnecessarily place workers at risk. Practitioners should bear in mind that clearances are just one objective of many.

Utility vegetation managers should establish and document appropriate clearance distances or vegetation heights or densities to be achieved at the time of work. Following work, vegetation on the ROW should be maintained at a stocking level that meets program goals and maintenance objectives such as reducing electric-safety and service-reliability risks, ensuring environmental stewardship, and controlling costs. Stocking levels are a measure of canopy cover height, density, and species mix. The best practice is to remove incompatible plant species, encourage compatible vegetation, and assure, through ongoing monitoring and maintenance, that vegetation is not established in areas where it will come anywhere near MVCDs.

Wire-Border Zone Concept

The wire-border zone concept is a management philosophy that was initially proposed by William Niering in 1958. It was promoted by William C. Bramble and William R. Byrnes, who added it to a study begun in 1953 on

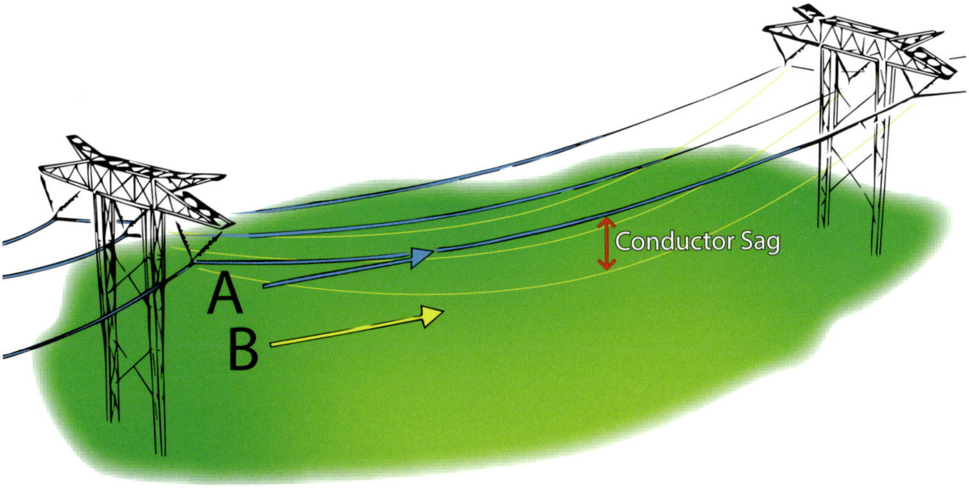

Figure 9. Line sag due to high ambient temperature, heavy electrical loads, or both. "A" is the position of the transmission line under low electrical loads and low-to-moderate ambient temperature. "B" is the maximum potential sag under highest engineered electrical loads and high ambient temperature. Lines that appear to have ample clearance from conductors under some conditions might still be exposed to cascading outages under other anticipated circumstances. Minimum vegetation clearance distances need to be established from the extremities of the engineered positional continuum.

the Pennsylvania State Game Lands 33 Research and Demonstration project. That research is still underway. The wire-border zone concept applies the principles of cover-type conversion to establish plant communities that do not have the genetic propensity to encroach on tolerance levels while providing a considerable margin of error.

The wire zone is traditionally described as the section of a utility transmission ROW between the outside phases, extending outward from the conductors to a distance specified by a utility vegetation manager. It is the section where lines are most exposed to sag, particularly during periods of high electric demand and high ambient temperature, which often go together. This is critical because the area directly under conductors presents the greatest risk of experiencing outages.

The wire zone is managed to develop low-growing plant communities dominated by grasses, herbs, and small shrubs. Low-growing shrubs are promoted in the wire zone as they complement the compatible cover type's capability of competing against tree invasion and enhance environmental stewardship. Low-growing plant communities also allow for ready access for inspection and facility maintenance. Establishing a wire zone over an entire ROW often misses the point and can create unnecessary expense through needless maintenance. However, woody-plant-free cover can be a legitimate maintenance objective, particularly along access roads associated with the ROW.

The border zone is the remainder of the ROW. It is managed to establish shrubs and trees with a maximum height that is well below specified tolerance levels. The border zone can provide a diversity of habitat for wildlife and serve as an ecotone between the wire zone and areas outside the ROW. The border zone is not only resistant to invasion by incompatible tree species but is also compatible with electric facilities.

Modification of the Wire-Border Zone Concept

Although the wire-border zone concept is regarded as a best practice, it often requires modification for several reasons:

- It does not address the need to keep the area around structures clear of woody vegetation.
- Right-of-way widths do not always have room for more than a relatively narrow border zone.

- There is no accommodation for the differences in line height and movement throughout a span.
- It does not account for topography.

A 2007 article by Benjamin Ballard, Kevin McLoughlin, and Christopher Nowak in the journal *Arboriculture & Urban Forestry* promotes enhancements to the wire-border zone concept. First, the authors observed that the area around the base of structures, including footings, should be kept free of woody vegetation. Keeping structures free of vegetation protects them from potential damage caused by woody plants, and fire, and allows access for inspection and maintenance. This space is called the exclusion zone.

The authors modified the border zone width to a more realistic representation of the ROW (Figure 10). Their border zone is narrow but still provides a transition as herbaceous species and lower growing shrubs in the wire zone blend into taller shrubs and small trees toward the ROW edge. The border zone progresses, in turn, toward the plant community outside the ROW. Species selection depends on site factors and maintenance objectives. However, no species should be promoted that have the genetic propensity to encroach on MVCDs.

Ballard and his colleagues observed that ground-to-conductor clearances are not uniform along transmission spans. So, they proposed that the wire-border zone concept be modified to account for variable wire height above ground. Considerations in adapting the concept include the line's maximum engineered movement (sag and sway), the potential lowest wire height off the

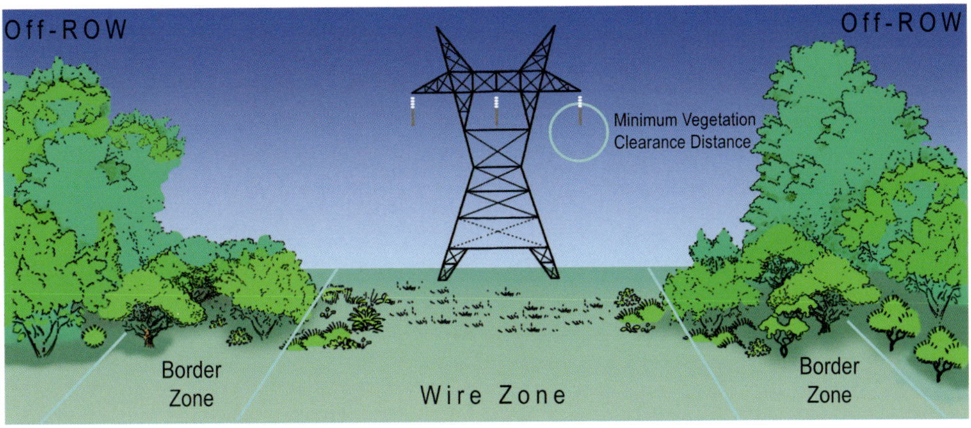

Figure 10. Modified wire-border zone (based on an article by Ballard, McLoughlin, and Nowak).

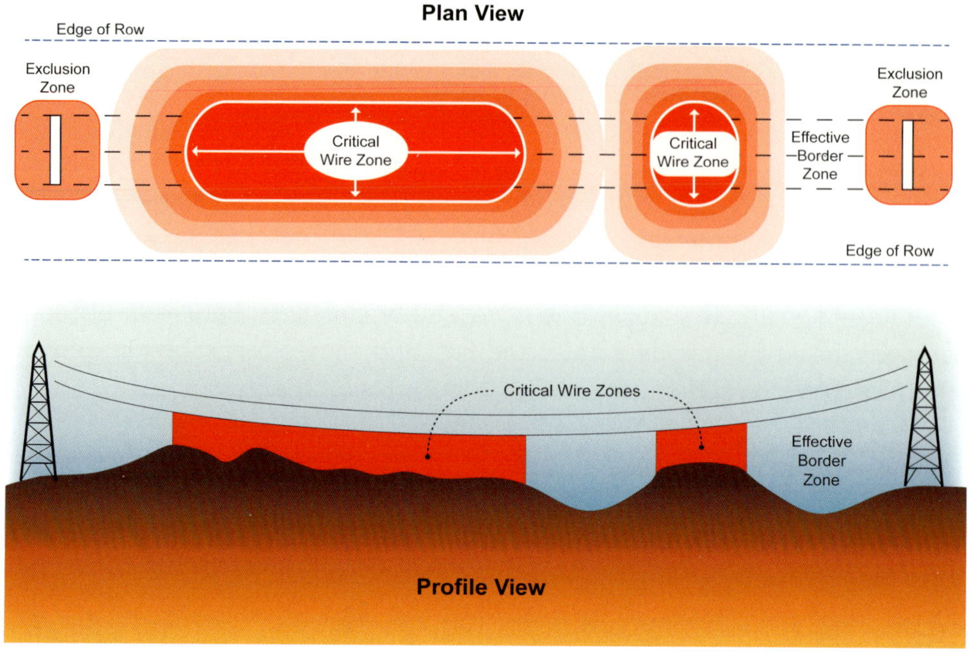

Figure 11. Plan and profile views of a transmission ROW modified to accommodate variations in wire height. This example assumes the structures are tall enough to accommodate an effective border zone community.

ground under normal designed operating conditions, access, structure height, stakeholder interest, environmental factors, wildlife, and other elements. On level terrain, ground-to-wire height is often lowest midspan, where line sag and sway are greatest. Those areas are designated as critical wire zones. Potential maximum movement gradually attenuates from midspan to the structures, to which conductors are fixed in place. In cases where towers are tall enough or wire is sufficiently high off the ground (as might be the case over swales and depressions), a typical border zone community can be promoted throughout the ROW between the critical wire and exclusion zones. Since the area is not limited to the borders, it is termed an effective border zone (Figure 11). In cases where structures are shorter, the critical wire zone might have to extend the entire span length, as is customarily depicted.

The principle can be broadened to areas where the conductors are high enough off the ground to accommodate timber. An example is where transmission towers are constructed on mountaintops or hilltops and power lines span over deep valleys or canyons. In some of these cases, sections of ground-to-conductor clearance is sufficiently high to permit timber species without risk of encroaching on MVCDs. This area can be considered a topographical zone.

An effective border zone might be established where ground-to-wire height decreases as the span nears the towers (Figure 12). If the structure height is too low to accommodate an effective border zone, a critical wire zone might have to be established as the conductors approach the exclusion zone.

These modifications have advantages. For instance, leaving trees or shrubs in valleys and canyons where lines are elevated provides environmental benefits. Streams often course through these low areas. Managing for topographical or effective border zones in valleys, canyon bottoms, or other low-lying areas helps shelter this valuable habitat and, if carefully selected, can be maintained without threat to the transmission lines. Furthermore, topographical zones can contribute to wildlife corridors. They also have economic benefits insofar as unnecessarily removing trees or other woody vegetation is a waste of money.

The wire-border zone concept may also be modified to accommodate side slope where ground-to-conductor clearance is higher on one side of the ROW than on the other. In such cases, border zones might be expanded on the uphill side of the ROW and reduced on the other to allow for topographical variance and wire height (Figure 13).

Decisions about where to establish topographical, effective border, critical wire, and exclusion zones should be made by vegetation managers and based on site and vegetative conditions. Gathering the necessary data to establish a modified wire-border zone can be facilitated with lidar or other technology. Such determinations should be based on vegetative and site conditions and on stakeholder interests, action thresholds, tolerance levels, line type

Figure 12. Profile of a topographical zone over deep canyons or valleys.

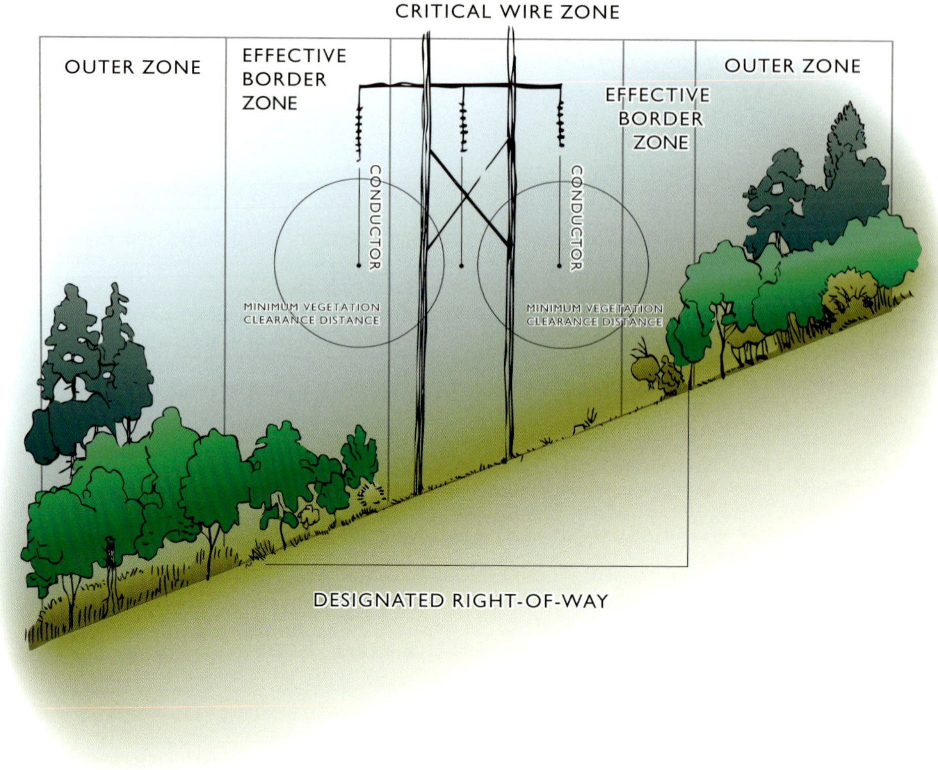

Figure 13. Wire-border zone modified to accommodate side slope.

and criticality, engineered wire movement, environmental considerations, economics, and other factors.

Wire-Border Zone in Fire-Prone Areas

Strict adherence to wire-border zone methodology may be inappropriate in some fire-adapted ecosystems, where border zone establishment is often best discouraged out of concern it could provide ladder fuels to adjacent lands. In these and other cases, management objectives could call for perennial meadow or prairie plant communities throughout the ROW.

Defensible space can also be maintained by leveraging off-ROW areas along power corridors or around hydroelectric plants and other facilities. Fuel levels can be reduced by applying a variety of control methods such as thinning trees, removing flammable shrub species, masticating deadwood, and other techniques. Horizontal and vertical spacing of vegetation can reduce the intensity of potential fires. To protect structures from fire, exclusion zones may be intensified to include clearing all vegetation to bare ground to a distance determined to be appropriate for the site. Exclusion zones maintained to bare ground can also serve as a fire-preventive measure against structures that have equipment that is prone to spark.

Engineering Solutions

While they are not vegetation control methods, engineering solutions can provide relief from vegetation–power line conflicts. They can include relocating, reconstructing, or burying lines. The disadvantage of engineering solutions is that they are often unaffordable for adjacent property owners or are not cost effective for utilities and their ratepayers. They can also have detrimental environmental impacts if inappropriately applied.

Appendix 2

Pipeline IVM

Many program goals for managing pipeline ROWs are similar to those of electric utility ROWs. Examples include safety, route identification, testing, encroachments, and maintenance and inspection, particularly aerial and ground patrol needed for leak detection. Route identification is important for underground facilities, which are pinpointed by aboveground markers or valves and measuring stations adjacent to the pipeline, which can be overgrown by unmanaged vegetation.

Vegetation can also hinder access to and maintenance of pipelines. Importantly, vegetation can obstruct underground pipelines, making it impossible to detect leaks from the ground or air. So it is often best to select plant species that are sensitive to gas, as pockets of dead or damaged plants can help identify gas leaks. Local botanists should be consulted for types of species that are sensitive to natural gas exposure. Tree roots may damage underground pipelines by compromising the integrity of the coating of some lines.

FERC Pipeline Requirements

In the United States, the Federal Energy Regulatory Commission (FERC) has some environmental requirements that pertain to vegetation management. They include:

- A 25-ft (7.5-m) vegetated buffer strip upland from the high-water mark on riparian areas
- A 10-ft-wide (3-m-wide) clear zone over the pipe
- Removal of trees that could impact the pipe out to 15 ft (4.5 m) from center
- Prohibition of pesticides within 100 ft (30 m) of bodies of water, unless permitted by a competent regulatory authority
- Prohibition of mowing or clearing from April 15 to August 1 to protect migratory bird nesting
- Exclusion of invasive species if they are not already abundant in adjacent wetlands

In the United States, the Pipeline and Hazardous Materials Safety Administration (PHMSA) under the United States Department of Transportation enforces the Hazardous Liquids Safety Act. The act requires unusually sensitive areas to be inspected at least 26 times a year but at no more than three-week intervals. Locations of drinking water or ecological resources, which could be damaged by hazardous liquid pipeline release, are considered unusually sensitive areas.

Pipe-Border Zone Concept

The pipe-border zone concept is a modification of the wire-border zone concept in electric IVM (Figure 14). It divides a pipeline ROW into a pipe zone and a border zone. The pipe zone is the area over the pipe, extending outward toward the ROW edge a specified distance. In the pipe zone, native forbs, grasses, and low-growing shrubs may be encouraged. A narrow path is often mowed directly over the pipe to facilitate access for testing and maintenance. Dense, low-growing, gas-sensitive cover could also be introduced into the pipe zone. The outer portion of the ROW is the border zone where shrubs, low-growing trees, and other compatible vegetation can be managed, provided the vegetation will not interfere with maintenance or pipe integrity. Species selected for the border zone should not interfere with access for inspection or maintenance nor cause root obstruction to the pipeline.

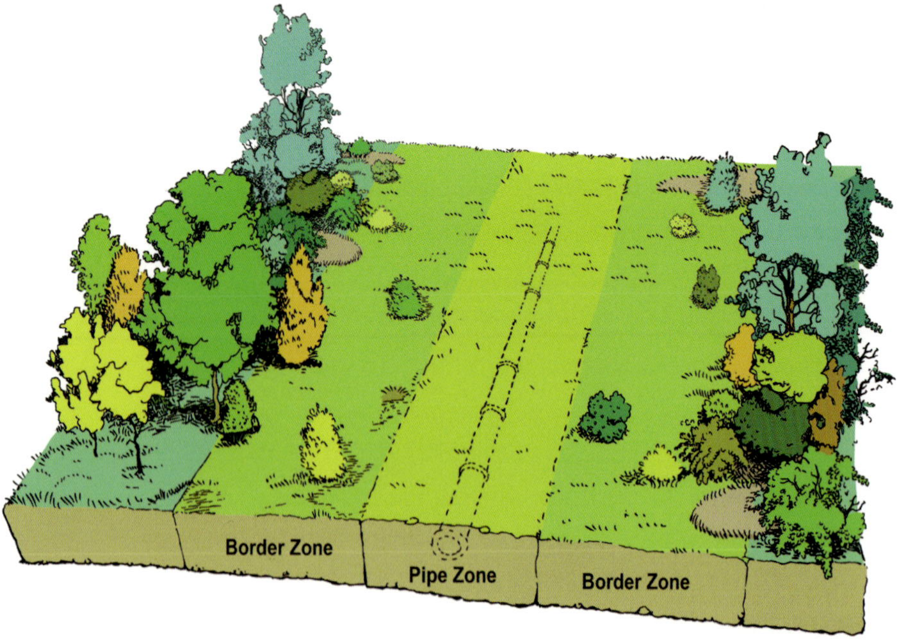

Figure 14. Pipe-border zone concept.

Appendix 3

Railway IVM

Railway IVM objectives include ensuring safety, protecting infrastructure and equipment (including ballast, tracks, communication lines, switches, and signals), minimizing trackside fire risk, and facilitating inspection of the railroad ROW. General principles of IVM presented earlier apply to railway IVM, as they do to other applications. Safety objectives can include maintaining lines of sight along the ROW for condition inspection and sign and signal visibility (including nighttime reflective conspicuity), as well as securing visibility for motorists at public railroad crossings, including those that are unregulated. Railway IVM helps protect train crews while they are entraining, detraining, switching, and performing other responsibilities. Fire risk can originate from brake-caused sparks, overheated bearings, and wheels.

Railroad track inspection and maintenance is essential for safe operation of every vehicle on the line. To have complete access to the tracks and surrounding areas, that area must be clear from trees and vegetation that could impede maintenance machinery. A clear ROW allows easier access for vehicles, which is important because not everything can be repaired and maintained while on track. Protecting infrastructure can involve managing risk of trees that could fall and strike trains, block the tracks, or damage switches and other equipment.

Railway Vegetation Management Zones

Railway vegetation management is often conducted in zones like those used by electric utilities, pipelines, and roadways. The zones consist of a ballast zone, an inner zone, and an outer zone (Figure 15). The ballast zone is the railway bed made of crushed rock, which supports the track and the weight of trains. The inner zone runs directly adjacent to the ballast zone and is managed in low-growing forbs and grasses, like the wire zone in electric ROWs or the pipe zone in pipeline ROWs. The outer zone is at the ROW edge and serves as a transition to adjacent lands. The outer zone is analogous to the border zone in electric IVM and is comprised of shrubs and short-statured trees.

The ballast zone must be kept completely free of vegetation, primarily to maintain structural integrity, which is required for stable track. Roots of vegetation retain fine soil particles and organic matter, which not only impedes drainage, potentially undermining the ballast, but also improves growing conditions for more vegetation. That further compromises the ballast, and conditions could degrade until the ballast is incapable of supporting the track in proper alignment, risking train derailment. Vegetation in the ballast zone can also obstruct laser-guided track alignment and automated infrared scanning for inspection of equipment such as hot boxes, flat-wheel and lateral load detectors, crossing signals, and power switches. Vegetation can also create slippery conditions that compromise traction and braking. Moreover, vegetation in the ballast zone potentially threatens electric lines that serve track heaters, signals, and switches.

The inner zone is maintained in a low-growing plant community to preserve visibility of signs and signals, minimize fire risk, provide visibility at curves, and protect the safety of train- and track-maintenance staff. Drainage ditches or swales are usually located in the inner ROW. The outer zone is often subject to selective vegetation management to control overhang, keep electric equipment clear, protect the tracks from tree blowdowns, keep road and

Figure 15. Railway IVM zones.

pedestrian crossings unobstructed, mitigate unauthorized access to the ROW, and protect normal functioning of equipment, including trains. Overgrowth of vegetation along the railroad tracks can cause false signal indications or disrupt communications vital to safe train operation. Both cab signaling and track signaling require a clear line to the broadcasting unit.

Appendix 4

Roadway IVM

Roadway IVM goals include safety, access, corridor continuity, protection of infrastructure, environmental stewardship, aesthetics, maintenance of buffers or shade for pedestrians, noise abatement, stress reduction, long-term maintenance efficiency, and others. Roadway IVM is consistent with vegetation management for other purposes.

Safety objectives can consist of maintaining sight lines along roads and at intersections, curves, and corners so motorists can see traffic control signs or approaching vehicles. Lines of sight can also be designed to influence traffic in situations like roundabouts, for example, where screening is commonly established in the center island to focus drivers' attention on the approaching traffic-direction shift. The same principle can be applied at "T" intersections where vegetation can help notify motorists to prepare to stop and turn. Vegetation alerts drivers of roadway boundaries and attenuates highway glare. Roadside features also provide motion cues that help motorists perceive their speed. Trees, fence posts, signs, and other objects create peripheral motion and flickering that may mitigate driver fatigue that can develop on long, otherwise monotonous stretches of road.

Wherever possible, a clear zone should be maintained along roadway edges to provide a smooth transition from the shoulder through the ditch line for errant vehicles. Objects such as tree stumps, boulders, or other obstructions can be hazardous to vehicles that leave the road, so they need to be removed. Trees should also be maintained to ensure sign visibility and provide unobstructed lines of sight in such areas as embankments, hills, curves, dips, and residential neighborhoods. Enough horizontal and side clearance from trees should be maintained to allow safe passage of trucks and other vehicles. High-risk trees or overhanging branches that can fall into roads need to be mitigated using principles outlined in this BMP. Vegetation can also be managed to reduce shade that can cause prolonged roadway icing or prevent the road from drying after precipitation events.

Roadside vegetation should be managed to mitigate wind and snowdrifts. Agencies responsible for areas subject to heavy snow need to manage drifts and maintain a zone for snow storage out of the mainline. Roadway IVM can also be applied to create defensible space along roadways to protect infrastructure and motorists from wildfire. Defensible space can alleviate the likelihood of wildfire originating from vehicles or motorists, reduce direct impacts to highway assets when wildfires do occur, and maximize traffic flow for emergency operations.

Protecting infrastructure is another objective of roadway IVM. Maintenance crews need unrestricted access to inspect traffic signal control boxes and ancillary structures such as guardrails, culverts, and bridges. Inspections focus on intruding roots, blocked drains, or other unforeseen conditions that could adversely affect roadway performance or design features. Moreover, roads built on levees and dams require inspection and corrective maintenance to ensure these facilities are not undermined. Vulnerable infrastructure can include pavement, retaining walls, or any support footing. Culverts and ditches also require periodic vegetation management to ensure proper drainage.

Roadway IVM Zones

Departments of transportation manage roadside ROWs in zones like pipelines, railways, and those administered by electric utilities. In the case of roadside ROWs, the zones are designed for motorists' safety as well as other objectives such as corridor continuity and environmental stewardship. Roadside ROWs are often divided into three sections (Figure 16). Zone 1 is the pavement or shoulder, which is either paved or gravel. Shoulders are often exploited by undesirable weedy species that tolerate poor growing conditions, so they need to be treated to be kept free of vegetation. Zone 1 is often managed by broadcasting nonselective and preemergent herbicides (see Appendix 7 for a discussion of selectivity and preemergent herbicides). Zone 2 is usually managed to be dominated by grasses and maintained through frequent mowing and broadcast herbicide applications. This zone is intended to provide a safe area for vehicles to recover if they exit a roadway. Zone 3 is often an area designed for drainage. It may contain riparian or other plants that are adapted to standing water for at least part of the year. This is the ROW section with greatest opportunity for cover-type conversion. Zone 3 can be a visual and environmental buffer and a resource conservation belt that merges into neighboring land. It may accommodate a transition of shrubs and small trees into the adjacent landscape (much like the border zone of electric utility corridors).

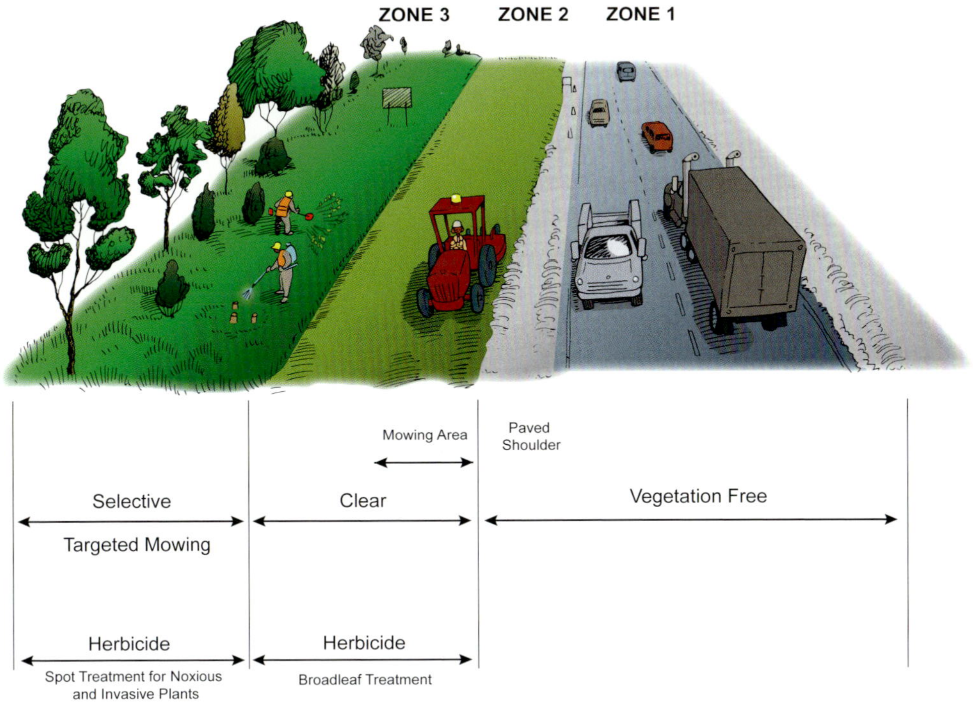

Figure 16. Roadway IVM zones.

Road ROWs

A road ROW is an area dedicated for traffic lanes, shoulders, ditches, and a safe area for vehicles to exit the roadway. Road ROWs are rarely owned in fee by the government, but they enable roadway management on property owned by others. Older roadways, such as those that progressed from an old wagon trail to a highway, might not have dedicated ROWs. Many roads have boundary fences, monuments, or delineators placed along the ROW to serve as guidance, not only for ownership but also for future construction and engineering purposes. The ROW is usually bordered by utility poles, as many jurisdictions have legislation enabling power and other utilities to occupy the road ROW for low or no cost. However, the presence of utility poles or other linear boundary markers may not accurately represent the true boundaries of the ROW. Roadway easements allow the use of another's property. There are specialty easements, such as cropping easements that enable harvesting hay and other activities.

Encroachments

An encroachment is an unauthorized building or object in the ROW. Examples of encroachments can be driveways, fences, buildings, flagpoles, advertising billboards, retaining walls, constructed water features, portable message signs, illegally parked vehicles, roadside memorials, or campers. Landscaping and agricultural production are also common encroachments on ROWs. Mailboxes are encroachments but ones that are usually allowed.

Appendix 5

Environmental Stewardship IVM

Vast reaches of North America were once dominated by prairies, meadows, glades, savannas, pine barrens, and similar plant communities. Due to pressures from agriculture, development, natural resource use, preference for trees, fire suppression, climate change, and other factors, much of this habitat has been reduced to pocket remnants, and many wildlife species are in decline as a result. Rights-of-way offer areas of opportunity that can be managed to restore this dwindling habitat while also accommodating the infrastructure located on them.

Stewardship on ROWs does not have to be complicated beyond fundamental IVM best management practices. It requires making environmental stewardship a management goal. As with any other IVM intention, environmental objectives guide action and influence data collection to apply to program assessment, scheduling, control method selection, progress, and quality control. Examples of wildlife and habitat stewardship goals include:

- Controlling aggressive non-native plant species
- Creating pollinator habitat by promoting plant species that produce high levels of nectar and pollen
- Encouraging specific wildlife populations
- Enhancing compatible habitat for biodiversity
- Establishing grasslands suited to ground nesting
- Improving wildlife corridors
- Promoting climate change adaptations
- Providing a diverse ecotone in the border zone

Stewardship activities are generally voluntary in nature and therefore can be paused, downsized, or reshaped with minimal concerns from regulatory pressures. The scope is flexible; it can extend across an entire system, focus on a region, or be relevant to a small patch of land. The size of the land available to a project can drive objectives because different species require various size territories. For example, small mammals and herptiles, pollinators, and some songbirds can often prosper in small, even disconnected pockets, while animals like bison, grassland birds, waterfowl, and predators frequently require large contiguous tracts.

Wildlife Habitat

Promoting wildlife habitat has been a primary goal of IVM since wildlife biologists and ecologists first developed cover-type conversion. The technique is used to manage vegetation to promote early successional habitat while inhibiting tall-growing woody species. It transforms ROWs to high-value assets for promoting biodiversity, conservation, and sustainability.

Successful IVM habitat planning must recognize the requirements of the wildlife being encouraged. Practitioners should be aware that promoting habitat for one species will often hamper another. For example, managing for grassland birds will displace forest species. Surveying to determine what wildlife is present can aid decision-making regarding which type of habitat will be most desirable in a project area. Developing suitable habitat requires an understanding of the target species' food, water, and shelter requirements, and how those factors are interrelated in the environment. It is important to recognize which plant species provide nutrition during periods of high energy demand for animal species being promoted. Such periods include breeding, lactation, or premigration. For example, flowering plant bloom times vary by the species and their seeds or berries develop at different times as well (phenology). Selecting vegetation to stagger the timing of such processes lengthens the period these resources can be used as food for target animals. An understanding of such seasonality can be leveraged to optimize habitat. An important consideration in that regard is developing diversity in native plant species. Diversity encourages a broad array of wildlife not only by meeting nutritional needs during different seasons but also by supporting shelter requirements. A diverse plant community also contributes habitat for mammals, pollinators, and insects, which in turn serve as a food source for many birds, herptiles, and bats in addition to many other species.

Corridors and Connectivity

Linear ROWs can both divide and connect habitat. They can be leveraged to provide contiguous tracts of habitat that can serve as connecting features in an otherwise fragmented landscape. They also can serve as seasonal movement corridors for migratory species such as mammals, birds, and insects. IVM can create a path between two patches of specific habitat (both on and off the

ROW). Animals may use this path to expand their range so they can more easily meet their needs. Online databases are available with information on species that are likely to make use of the ROW.

Examples of IVM activities supporting connectivity include:

- Avoiding disturbances during sensitive activities (breeding and rearing periods, for example)
- Leaving brush in the wire or pipe zones every couple of hundred yards (meters) that can be used as escape cover
- Monitoring IVM activities on the ROW segment that serves as a connector to resemble the conditions of the isolated patches as closely as possible
- Providing natural cover adapted to species likely to use the ROW segment as a connector
- Reducing or removing spatial barriers for targeted species, often around maintenance of access roads
- Understanding the species utilizing ROW segments

ROWs can be impediments to connectivity. Road and railway vegetation managers should be mindful of the potential risks to humans and animals of attracting wildlife to transportation corridors. The field of road ecology works through site evaluations and wildlife behavior surveys to identify ways to accommodate the needs of wildlife while reducing the likelihood of vehicle collisions with them. Examples of road ecology initiatives include building specialized bridges and tunnels to provide safe crossing points for wildlife.

Climate Change Adaptation

Vegetation management can help ecological communities adapt to the effect of climate change by contributing to carbon capture and other climate-related goals. However, the impact is usually minimal in a global context. Nevertheless, IVM can be applied seamlessly to climate change adaptation by making it part of management goals and maintenance objectives. IVM considerations attendant to climate change include fluctuations in the length of growing seasons and migration timelines, elevated fire and flooding risks, invasive species spread, and other factors. Some of these factors can be applied to linear corridors as described by the Wildlife Habitat Council (WHC) in Table 8.

Table 8. Recommended implementation of climate change adaptation strategies for ROW management (from the Wildlife Habitat Council).

Recommended adaptation strategies	Implementation of adaptation strategies for ROWs
Establish native plant communities on ROWs	New ROWs can be seeded with native seed mixes, existing corridors can be enhanced by controlling for incompatible invasive species, and specific spans of the ROW can be targeted for restoration.
Create wildlife corridors/ habitat linkages	Segments of the ROW can be managed to mimic the objectives and management techniques of high-quality adjacent off-ROW patches, creating continuous habitat. Linkages can also be created on ROWs by minimizing obstacles and drastic changes in vegetation communities.
Create/restore quality habitats	New ROWs can be seeded using native seed mixes, and existing ROWs can be enhanced by overseeding with native seeds or by planting natives.
Translocate species	ROW managers can assist in the movement of species (plants and animals) by partnering with their local wildlife agency to relocate the species on their land.
Create/manage buffer zones	ROW can be narrowed in length and, by working with neighboring properties, ROW can be managed in a way consistent with neighboring partners.
Manage invasive species	ROW can be monitored for invasive species and pests; ROW managers can also track the movement of invasive pests and diseases in an area and, by working with neighboring landowners, can make coordinated efforts to prevent the spread of invasive species and pests.
Utilize seed mixes that are more adaptive to climatic extremes	ROW managers can select fire- and drought-tolerant plants to lower the risk of fire damage to assets and species on the ROW.
Manage for shifting northern boundary of species ranges	ROW can be seeded with mixes that include grasses and forbs that are found in the lower range of the ROW for the whole area that falls within that range, assisting in the movement of species as the climate forces species to shift their ranges northward.
Recategorize invasive species selected for management control (fugitive species)	ROW managers should monitor invasive species that have moved into a new area; however, if the species is providing an unmet need in the habitat, caution should be exercised before outright removal.
Reduce non-climate stressors	ROW managers can reduce non-climate stressors by managing invasive species, increasing soil health using lime or fertilizers, and preventing habitat fragmentation by managing lands close together in a similar way.
Increase biodiversity on ROWs	ROWs can be seeded with a variety of native species with different blooming times and with varieties that meet different and changing habitat/species needs. Land that is diverse in structure and species composition is more resilient to change.
Educate to raise awareness about climate change	Joining committees and regional groups and leading or creating opportunities to increase employees' and community members' awareness of climate change and adaptation tactics being used can help build public support for these activities on ROWs.

Table 8. *(continued)*

Recommended adaptation strategies	Implementation of adaptation strategies for ROWs
Educate for building capacity in deployment of adaptation tactics	Contractors can be trained to recognize native communities and invasive species to implement adaptation techniques. Working with employees and contractors on proper seed-mix selection can lead to more effective management of ROWs and reduce costs associated with fixing ineffectively deployed techniques.
Research/monitor the impacts of climate change on ecosystems	ROWs can be managed using the techniques mentioned above and efforts should be made to actively publish research on test plots associated with climate change adaptation strategies.
Increase the carbon sequestration potential of ROWs to contribute to climate change mitigation efforts	ROWs can be managed for healthy plant communities, which sequester carbon.

Partnerships and Resources for Environmental Stewardship

Comprehensive environmental stewardship is outside the realm of expertise of many IVM managers. Fortunately, there are several nonprofit and academic organizations to which practitioners can turn for support of stewardship goals. They include Ducks Unlimited, the National Audubon Society, the Wild Turkey Federation, the Rights-of-Way as Habitat Working Group (RHWG), the UAA, Trout Unlimited, the Wildlife Habitat Council, the Natural Resource Defense Council, and the Wildlands Network, among many others. The WHC, RHWG, UAA, and Wildlands Network specifically concentrate on environmental ROWs. Readers are directed to these organizations' websites for more detailed information.

The WHC's guidance recommends four project categories: habitat, species management, education and awareness, and other options. Habitat projects center around conserving, protecting, and restoring habitats. Species management projects are designed according to the needs of targeted wildlife species. The objective of education and awareness projects is to improve the understanding, awareness, and proficiency related to conservation and the environment. Other options are specialized projects that add value to conservation.

The RHWG is headquartered in the University of Illinois at Chicago and is active in Canada and the United States. They collaborate with arborists, biologists, educators, engineers, environmentalists, foresters, lawyers, and other stakeholders in promoting habitat improvement on ROWs and other working landscapes. In total, the RHWG is aligned with more than 200 academic, corporate, governmental, and nonprofit organizations from across North America. Industrial enterprises include the gas, electric, and railroad companies and departments of transportation. The RHWG specializes as a source of information on promoting pollinator habitats and healthy ecosystems along ROWs.

The UAA values environmental stewardship and understands that excellence depends on managing ROWs as ecosystems. It maintains libraries of IVM and environmental resources. It also promotes ROW stewardship accreditation and sponsors this BMP.

The Wildlands Network is a North American conservation organization dedicated to promoting wildlife. Since roadways and other artificial obstructions block wildlife passage, the Network encourages development of wildlife corridors to protect and restore native species habitats. Animals that benefit from corridors are many and include cougars (*Puma concolor*), pronghorns (*Antilocapra americana*), grizzly bears (*Ursus arctos horribilis*), deer (*Odocoileus* spp.), migratory birds, monarch butterflies (*Danaus plexippus*), and other pollinators.

Appendix 6

Soil Health

Healthy soils are essential for thriving plant communities, which are central to effective IVM. This appendix provides an overview of soils and how they affect IVM. Readers interested in further information are encouraged to consult Best Management Practices: *Soil Management* (ANSI A300, Part 2), *Root Management* (ANSI A300, Part 8), and Chapter 3 (Soil Science) of the ISA's *Arborists' Certification Study Guide*, among other resources.

Soil is a product of the environment. It developed over long periods through mineral weathering, climate, topography, and the influence of organisms living in and on it. Soil consists of solid, liquid, and gas phases. Each phase has its own importance and impact on plant health. The solid phase has inorganic and organic constituents. The inorganic faction is mineral material derived from surface rock substrate eroded over eons. The organic component is made up of plant and animal remains, vegetative litter, and excretory products in various stages of decay. The liquid phase is also called the soil solution. It is water with dissolved elements and other substances. The gas phase is the soil's atmosphere, mainly found in macropores. All three phases are necessary to support the soil ecosystem.

Soil Structure

Soil structure describes the organization of soil solids, including aggregate formation and the pore space between them. Aggregates are clumps of solid-phase components. Aggregates determine the organization of micro- and macropores, both of which are essential for plant growth. Micropores retain water but little air. Macropores are too large to hold water but do hold air. Healthy soils have a balance of both so that water and air are available for the optimal soil ecosystem necessary to support a thriving plant community.

Soil structure development occurs most readily near the surface where the effects of organic matter, root activity, and freezing and thawing are most concentrated. Roots and ice develop soil structure by expanding in the pores, wedging the soil apart, and compressing particles into aggregates. Burrowing animals, particularly earthworms, also contribute to structure. These processes increase the ratio of macropores to micropores. Poorly structured soils, including compacted soils, may not have enough large pores to sustain a healthy soil ecosystem, which will compromise the vitality of a plant community and, in extreme cases, plant survival.

Organic Matter

Organic matter comprises a vital part of healthy soil's solid phase. Fallen leaves and other plant matter accumulate as a distinctive layer on top of soil and incorporate into it as they decompose. Vegetative litter forms an insulating mat on the soil surface that protects against extremes in temperature and moisture fluctuation. It also prevents erosion and facilitates water percolation and infiltration.

The organic layer is an area of intense biological activity because it is utilized as food by soil organisms, mostly microorganisms. Decomposed plant organic matter together with the remains of microorganisms becomes humus, a dark-colored, submicroscopic material. Humus enhances cation exchange and water-holding capacities and contributes gumlike, binding substances that function in building soil structure. Moreover, as organic matter is broken down, essential elements, notably nitrogen, phosphorous, and sulfur, are released into the soil. Higher levels of soil organic matter benefit plant vitality in all soil types. For example, organic matter adds water-holding capacity to sand, from which water easily drains, and improves soil structure with its attendant increased pore space in fine soils like clay, which often readily hold water at the expense of air.

Rhizosphere

Plants blend with rather than grow in the soil, forming a synergistic relationship. They influence soil characteristics, which in turn affect plant growth and vitality. That interaction occurs in the rhizosphere, which is an ecosystem that includes the soil and plant roots (Figure 17). The organisms involved range in size from small, one-celled bacteria, algae, and protozoa, to fungi, nematodes, and more complex micro-arthropods, to macroorganisms such as earthworms, insects, small vertebrates, and plants. Plant roots produce carbohydrates, amino acids, and other compounds that nourish microorganisms. In return, microorganisms stimulate or repress other soil organisms, including suppressing pathogens, and help plants absorb water and essential elements. In the soil ecosystem, microorganisms serve as a food source for mites, nematodes, springtails, and other microfauna that in turn are prey for spiders, centipedes, and other insects, which are a food source for larger animals such as birds, ground squirrels, and others. As organisms digest complex materials or consume other organisms, essential elements are converted from one form to another and are made available to plants and other soil organisms. So, plants depend on the soil ecosystem for their essential elements. The point is that healthy soil contributes to thriving plant communities, and vegetation managers should be mindful of how to promote optimum soil conditions to meet their objectives.

Human Activity

While mostly natural conditions create soils, human activity can disrupt and alter soil characteristics. For example, construction or other development often removes the organic layer, which decreases biological activity, hampers soil structure development, and compromises soil gas and liquid phases and, by extension, plant access to oxygen and water. It also compromises the rhizosphere, impeding elemental cycling and mycorrhizal activity.

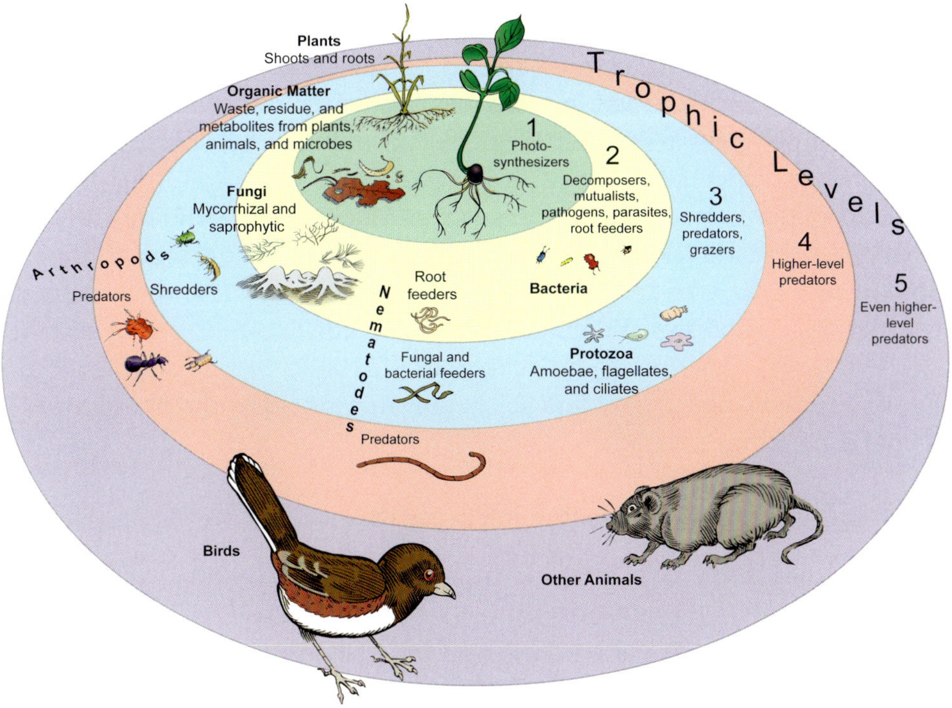

Figure 17. Organisms and organic matter make up the soil food web. This diagram shows a series of energy and essential element conversions as organisms and organic matter at lower trophic levels are consumed by organisms at higher levels.

Compaction may be another negative consequence of human activity. Compaction can be caused by vehicles, construction equipment, or heavy pedestrian traffic. It reduces total pore space along with the proportion of macropores to micropores and drives oxygen out of the soil. Loams and other soils with a variety of particle sizes can be particularly vulnerable to compaction because small particles are pressed into the large pores between coarse particles. Soils do not readily recover from structural damage, since structure takes a long time to develop. Reduced pore volume restricts aeration, drainage, and root penetration and favors shallow-rooted plants such as many weed species. The survival strategy of these species is to grow and reproduce quickly, making them undesirable on many ROWs.

Human activity can also lead to crusting or erosion, particularly when plants and organic matter are removed from the soil surface. Crusts can result in altered drainage, elevated pH, and a physical barrier to water infiltration, gas exchange, and seedling penetration. All these factors may harm root growth and the health of a plant community.

Mixing occurs when soil is scraped, stockpiled, and re-spread. In some cases, topsoil or fill is hauled in from off-site to be spread on top of existing soil. Scraping destroys soil profiles in a manner analogous to soil erosion. Mixing creates abrupt changes in soil texture, organic content, or bulk densities. These abrupt changes differ from the more gradual transitions often found under natural conditions and may compromise aeration, water-holding capacity, drainage, fertility, and root growth. For example, if very fine-textured topsoil is spread over a coarse-textured soil, a perched water table may result in the upper layer. Adhesive and cohesive forces in the fine-textured layer hold water tightly and may not readily release it. The underlying coarse-textured soil cannot draw water out of fine soil, and water is held by the fine-textured soil until it becomes saturated.

A soil texture triangle is presented in Figure 18.

Erosion

Erosion is soil surface abrasion by wind or water. It compromises healthy soil, resulting in compaction, poor soil structure, impeded drainage, pH alteration, reduced organic matter content, and low fertility, among other problems. It is caused when permanent cover vegetation is lacking and an area is exposed to wind or water, conditions that rarely occur in nondesert natural areas.

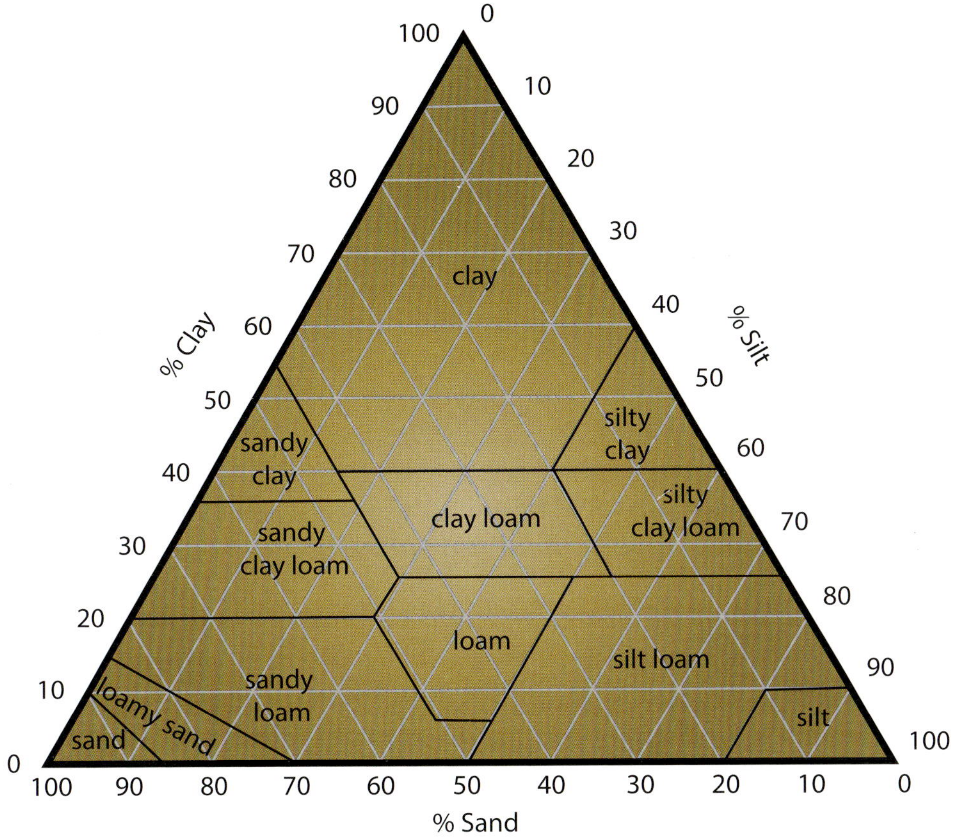

Figure 18. Soil texture triangle.

There are several types of erosion: splash, sheet, rill, channel (gully), and bank. Splash erosion is caused by raindrops that redistribute soil particles. The relocated particles create a crust on the surface, which impedes subsequent water infiltration. Sheet erosion is soil dislodged from uniformly smooth surfaces by raindrops and runoff water. It is displaced and deposited in low areas like the bottom of slopes. Rill erosion occurs from slight differences in elevation, which cause runoff to concentrate and form small but well-defined ruts or rivulets (rills). Channel erosion is advanced rill erosion where resulting ruts expand into well-defined gullies. Channel erosion is difficult to correct. Bank erosion is progressive scouring and undercutting of streambanks or constructed drainage channels. It is often caused by uncontrolled livestock access, poor maintenance, or failure to establish vegetative buffers along streams and waterways.

Mitigation

Soil analysis can help managers determine deficiencies in the soil that need to be mitigated. Factors that managers should consider include the soil pH; percent organic matter; soil texture; water-holding capacity; the amount of available nitrogen (N), phosphorous (P_2O_5), and potassium (K_2O); salinity; cation exchange capacity; and other factors.

Often problems caused by mismanaged soils can be overcome with site preparation. For example, surface compaction may be corrected by tilling, and poor drainage may be remedied by installing surface or subsurface drainage systems. If water is limiting, irrigation systems may be built. In some cases, existing soil may be replaced with a designed growing medium. Soil design attempts to re-create natural soil horizons that are suitable for plant growth. Readers interested in more information on designed soils, drainage, irrigation, and other pertinent issues may consult *Urban Soils: Applications and Practices* by Phillip Craul.

It is a best practice prior to construction to separate the humus layer, then work it back into the soil after the project is complete. If the humus layer has been removed and is unavailable, organic matter can be amended into the soil as a mitigation measure. Mulching the soil surface can also be beneficial.

Appendix 7
Herbicides

Herbicides are crucial to IVM, as they are generally the most economical and effective means of facilitating biological control. Their use can create public concerns, and improper application can result in unintended, negative consequences. To be used safely, they must be handled according to label directions and governmental regulations.

Toxicity, Site of Action, and Mode of Action

Toxicity is a substance's ability to injure a living system (plant, animal, or ecosystem) and refers to the extent of damage it can do. A fundamental principle in understanding toxicity is the interaction between exposure and dose. Exposure is the amount and frequency a living thing or environment comes into contact with a substance. Dose is the volume of chemical that is absorbed. In general, greater exposure means a higher dose. Herbicides used in IVM are designed to disrupt botanical processes. When used at the rates and frequency prescribed by their labels, the levels of exposure are not harmful to animals, including humans.

A herbicide's mode of action (MOA) is its botanical biochemical interference and occurs on a site of action (SOA). A MOA is how a herbicide works, while a SOA is where in a cell the herbicide controls a plant. SOAs determine a chemical's class. MOAs and SOAs of herbicides commonly used for IVM are presented in Table 9.

Managing Herbicide Resistance

Herbicide resistance is becoming increasingly prevalent, especially where herbicide is applied on short (e.g., annual or more frequent) intervals on herbaceous species. Herbicide resistance on woody plants is less likely due to typically longer application intervals. Yet, resistance impacts nearly every herbicide class. Best management practices have historically recommended rotating MOAs to reduce the likelihood of herbicide resistance developing within an area. However, better results have been obtained by mixing multiple herbicides with different SOAs in each tank and ensuring active-ingredient volumes are never below the minimum application rates.

Table 9. Modes of action (MOAs) of classes of commonly used IVM herbicides.

Chemical class (site of action)	Mode of action	Common IVM active ingredients[1]	Selective or nonselective
ALS or AHAS inhibitors	Inhibits key botanical enzymes necessary for creating essential amino acids	Imazapyr, metsulfuron-methyl, sulfometuron-methyl	Depends on active ingredient
Cellulose inhibitors	Inhibits cell-wall formation	Indaziflam	Nonselective (preemergent)
EPSP inhibitors	Inhibits key enzymes needed to create botanical proteins or biochemical pathways required for growth	Glyphosate	Nonselective, not soil active
Photosystem I inhibitors	Steals electrons from photosynthesis (photosystem I) and uses them to form compounds that destroy cell membranes and chlorophyll	Diquat	Nonselective
Photosystem II inhibitors	Stops photosynthesis so plants starve to death	Diuron, tebuthiuron, triazine	Nonselective
Potential nucleic acid inhibitors or nondescript MOA	Bud inhibitor	Fosamine	Selective to most non-woody species, not soil active
Synthetic auxins	Acts similarly to a natural plant growth regulator (auxin). Affects cell-wall plasticity and cell metabolism. Also stimulates accelerated cell division and plant growth in broadleaf plants and trees, which results in vascular tissue destruction.	2,4-D, aminocyclopyralid, aminopyralid dicamba, florpyrauxifen picloram, triclopyr	Selective

[1] There are often multiple trade names for individual active ingredients.

Water Considerations

Hard water can compromise herbicides that are weak acids, such as glyphosate and 2,4-D. Hard water usually has high pH, which reduces the effectiveness of weakly acidic herbicides. Elevated pH causes the acidic herbicide to partially dissociate because high cation concentrations (particularly sodium) found in hard water bind with the dissociated herbicide, weakening its effectiveness. The problem can be mitigated by adding adjuvants, like ammonium sulfate, to reduce the pH of a hard water–weak acid herbicide mix.

Turbid water (water contaminated with suspended solids, soil, or organic matter) can also cause problems. Suspended soil colloids can bind to and inactivate some active ingredients. Practitioners should find another water source if the water they have available is murky or discolored.

Adjuvants

Adjuvants are materials added to a mix or are part of the herbicide formulation that improves application characteristics or herbicide efficacy. Adjuvants can increase the range of conditions where specific herbicide formulations can be used effectively. They can also improve herbicide performance by facilitating absorption into plants. Surfactants are types of adjuvants that reduce the surface tension of herbicide-mix droplets so they spread out when applied to plants, which improves coverage, absorption, and translocation. Examples of adjuvants include methylated seed oil, crop oil concentrates, alkylphenol ethoxylate, alcohol ethoxylate, and ammonium sulfate.

Tree Growth Regulators

Tree growth regulators (TGRs) are substances that reduce plant growth. They can be helpful to slow growth of some trees that cannot be removed, though they are problematic because they have the genetic propensity to interfere with critical infrastructure. Slowing tree growth has been shown, in some cases, to reduce long-term costs by extending the length of maintenance intervals.

There are two common active ingredients in TGRs: flurprimidol and paclobutrazol. Flurprimidol is water soluble and is typically injected into tree trunks. Paclobutrazol in suspension is applied by soil injection or basal soil drench. TGRs not only slow shoot growth but also cambial expansion. Treated plants generally have deeper green-colored leaves that suggest elevated chlorophyll content compared to untreated plants. There is also evidence that root growth can be enhanced in some cases. Treated trees often are more tolerant of water stress, as they have increased root growth, reduced leaf surface, and thicker, more dense masses of hairs on some species. The latter two characteristics are drought adaptations. Paclobutrazol has been found to have fungicidal properties.

Selectivity

Selectivity refers to either the types of plants that a herbicide controls or the application technique used. Selective herbicides work only on certain types of plants, leaving others unharmed. On the other hand, nonselective herbicides control nearly all vegetation, but even these can be selectively applied to individual plants or groups of unwanted plants, leaving desirable species untouched.

Selectivity and herbicides can involve:

- Selective use of either selective or nonselective herbicides that only control targeted vegetation. Selective use is preferable unless target vegetation density is high.
- Selective applications that are used against specific plants or pockets of plants.
- Synthetic auxins, which are a class of selective herbicides that control broadleaved plants but do not harm grass species (when appropriately applied).
- Nonselective herbicides that control most of both broadleaved plants and grasses.
- Nonselective techniques that target areas rather than individual plants (see Herbicide Application Methods).
- Nonselective use of nonselective herbicides, which eliminates all treated plants in the application area.
- Nonselective use of selective herbicides, which controls treated plants in the treatment zone that are sensitive to the herbicide, without affecting plants with low sensitivity.

Pre- and Postemergent Herbicides

Pre- and postemergent herbicides can be used to advantage in meeting maintenance objectives. They can be applied alone or in combination. Preemergent herbicides help to inhibit germination and can reduce the ability of incompatible species to establish. These products help reduce seed banks of incompatible plants and are recommended as a best practice to control annuals. Preemergents are typically applied prior to the growing season when rain is likely to help the herbicide seep into the soil. They are commonly broadcasted on bare soil before germination to affect residual control. Bare-ground applications are appropriate on or around road shoulders, railway ballast zones, electrical substations, certain utility poles to protect against fire, and other areas where vegetation must be excluded.

Postemergent herbicides help control established plant populations and are commonly used to control perennial species. Used on their own, postemergent herbicides typically do not provide the benefit of residual control in the soil. However, a tank mix or a product with a combination of both pre- and postemergents can offer synergies of benefits. In either case, timing is critical. Postemergent herbicides are usually more effective when applied to plants prior to seed set.

Herbicide Application Methods

Herbicide application methods are categorized by the quantity and selectivity of herbicide used, application technique, character of the target, vegetation density, and site parameters. Dyes can be added to the herbicide mix to mark areas that have been treated. Treatments include individual plant and broadcast treatments. Herbicides are less effective when plants are drought stressed, as low water potential also slows translocation and metabolism, so herbicide movement in the plant will be impeded. Foliar applications might be particularly compromised because many plants respond to water stress by developing a thicker wax layer on leaf surfaces, which can impede herbicide absorption.

Individual Plant Treatment

Individual plant treatments are selective. They include cut-surface, cut-stump, basal, frill (hack-and-squirt), chemical side pruning, hydraulic (high-volume) foliar, low-volume foliar, and ultra-low-volume foliar. Because they are applied selectively, proper individual plant applications work well to avoid damage to surrounding sensitive or off-target plants. However, they are impractical on broad areas of tall or dense brush and sites dominated by undesirable species.

- Cut-surface herbicide treatments are applied to the freshly cut stump, concentrating on the cambium. These treatments are usually done with water-based formulations and must be applied as soon after cutting as possible so the herbicide is taken into the plant. Applications during spring sap flow should be avoided, as water-based herbicides will be diluted or forced off the cut surface by the sap.
- Cut-stump treatment is applying herbicide to the stump, bark, and root collar using an oil-based formulation for bark penetration. Cut-stump treatments may be applied hours, days, or even weeks after cutting. Applications can be made year round, even in cold weather regions (Figure 19).

Figure 19. Cut-stump treatment.

- Basal applications often use a herbicide in a vegetable or refined petroleum oil carrier. The technique is to apply the mix to the plant base, encircling stems and the root collar (Figure 20). The oil carrier penetrates the bark and carries the herbicide into the plant. Although basal applications can be made year round, dormant treatment is often desirable on deciduous plants, when they do not have foliage that can obstruct access to individual stems.
- Frill (hack-and-squirt) treatments consist of water-based herbicide application into wounds made in the stem. Frill treatments are especially useful against large incompatible trees that can be left standing for wildlife. Timing this application later in the summer or in early fall increases effectiveness. Efficacy is reduced during sap flow in the spring, as the herbicide solution is forced out of the cuts.
- Chemical side pruning is a technique where non-translocatable herbicides are applied to foliage of specific branches that could potentially interfere with objectives (like clearing electric facilities, traffic signs, buildings, vistas, or other objects), causing them to defoliate, die, and eventually be shed by the tree.
- Hydraulic (high-volume) foliar application uses a high-pressure system—including a tank, hose, and spray gun—to apply herbicide to incompatible plants across relatively broad areas (Figure 21).

- Low-volume foliar and ultra-low-volume foliar applications are done by treating target plants with specialized herbicide formulations and equipment, often with backpack sprayers. Low-volume foliar treatments are applied with flat fan nozzles. Low-volume foliar often leverages adjuvants to maximize control while minimizing herbicide active-ingredient volume. Ultra-low-volume foliar treatments use specially designed nozzles and a paraffin oil carrier that reduces spray drift and evaporation. The system requires only 7 to 10 droplets of herbicide mix on each leaf for control.

Figure 20. Basal application.

The amount of low-volume foliar herbicide applied depends on stocking levels, which is the percent coverage as determined by an interrelationship between height and density (Table 10). As a rule, the volume applied increases with greater average vegetation height at a particular density and with increasing density at a given height. Table 10 provides an example to illustrate stocking.

Broadcast Treatment

Broadcast treatment is nonselective control of all plants sensitive to a specific herbicide in a treatment area. Broadcasting is particularly useful against large infestations of incompatible vegetation (including invasive species). Broadcast treatments can provide a degree of selectivity with selective herbicides. For example, broadcast foliar treatments of synthetic auxins will control all the broadleaved plants in the target area. At the same time, with nonselective

Figure 21. Hydraulic (high-volume) foliar application.

Table 10. An example of a stocking matrix used to identify suitability and cost of vegetation maintenance treatments.

STOCKING			HEIGHT				
			short	*medium*	*tall*	*very tall*	*extra tall*
DENSITY			<3 feet <1 meter	<6 feet <2 meter	<10 feet <3 meter	<13 feet <4 meter	≥13 feet ≥4 meter
ultra-light	<50/acre	<125/hectare	5%	5%	10%	15%	20%
very light	<500/acre	<1250/hectare	10%	20%	30%	40%	50%
light	<1000/acre	<2500/hectare	30%	50%	80%	100%	100%
medium	<3000/acre	<7500/hectare	80%	100%	100%	100%	100%
heavy	≥3000/acre	≥7500/hectare	100%	100%	100%	100%	100%

technique, selective herbicides might not differentiate between compatible and incompatible plants that the chemical controls.

Broadcast techniques include bare-ground, broadcast-foliar, cut-stubble, and aerial applications (Table 11).

- Bare-ground treatments are used for clearing all plants in a prescribed area, such as in substations, around poles, in the railroad ballast zone, on roadway shoulders, and where vegetation is removed to protect infrastructure from fire. Bare-ground applications are often combinations of pre- and postemergent herbicide following mechanical vegetation removal.
- Broadcast-foliar applications target a broad area of incompatible species rather than individual plants or pockets of plants. Applications are used with calibrated spray equipment (such as a "radiarc" or "boom buster" guns). Broadcast-foliar applications can begin once stems have hardened off and plants are in full leaf and may continue throughout the growing season. Applications are often used in areas of high incompatible plant density or as a follow-up to physical control after cut vegetation has resprouted.
- Cut-stubble applications are made using either high-, low-, or ultra-low-volume broadcast treatments over areas that have just been masticated. Cut-stubble application utilizes soil-active herbicides that control root systems of potential root and stump sprouting species. Buffer zones should be established along ROW edges to ensure herbicides are not absorbed through the root systems of nontarget trees.
- Aerial applications are broadcast treatments made by helicopter (rotary wing), small airplane (fixed wing), or unmanned aerial systems (drones). Rotary wing aircraft provide accuracy because helicopters can hover, are more maneuverable, and can fly more slowly than airplanes. However, airplanes are less expensive to operate than helicopters. Unmanned aerial systems are smaller, less expensive, and more maneuverable than helicopters. They have the potential to provide the highest degree of aerial application selectivity, particularly when coupled with GIS software programmed to treat specific target areas.

Table 11. Herbicide treatment type and selectivity.

Application technique	Herbicide type: Selective herbicides (controls certain types of plants)	Herbicide type: Nonselective herbicides (controls all types of plants)
Selective application (targets individual plants)	Most selective. Controls only treated plants that are susceptible to the herbicide. • Basal • Cut-stump • Cut-surface • Frill (hack-and-squirt) • Hydraulic (high-volume) foliar • Low-volume foliar • Ultra-low-volume foliar • Side pruning	Intermediate selectivity. Controls all treated plants. • Basal • Cut-stump • Cut-surface • Frill (hack-and-squirt) • Hydraulic (high-volume) foliar • Injection • Low-volume foliar • Ultra-low-volume foliar
Nonselective application (targets an area)	Intermediate selectivity. Controls all plants in the treated area that are susceptible to the herbicide. • Aerial • Broadcast-foliar • Cut-stubble	Least selective. Controls all plants in treated area. • Aerial • Bare-ground • Broadcast-foliar • Cut-stubble

Closed Chain of Custody

Traditionally, herbicides have been distributed in concentrated forms in single-use, disposable containers. Spent containers need to be triple rinsed and disposed of. This requires handling open containers of concentrate on job sites for measuring, mixing, loading, and disposal. Advances in chemistry and application methods have significantly reduced the volume of herbicide needed in solutions. These advances have made it practical to adopt a closed chain of custody concept in which ready-to-use and diluted concentrate formulations are utilized in closed delivery systems (Figure 22), a practice that further protects the applicator and the environment.

The closed chain of custody concept includes herbicide shipping, distribution, storage, mixing, and record keeping. The process includes returning empty containers for refilling and reuse and involves four cycles:

- Container cycle: Returning, refilling, and reusing supply containers
- Integrity cycle: Closed connections at the transfer points between supply containers, mix tanks, and application equipment
- Documentation cycle: Container tracking system that establishes an auditable record documenting movement of herbicides and containers

Figure 22. Closed chain of custody.

- Herbicide cycle: Use of customer blends containing the required active ingredient and adjuvants

The UAA and ISA have produced best management practices for closed chain of custody for herbicides in the utility vegetation management industry. Interested readers are encouraged to consult these best practices for further information on the subject.

While not closed chain of custody, one-dump containers are noteworthy. One-dump containers are filled with specified herbicide blends, including adjuvants, "dumped" into a tank, and returned to be refilled. They are used on small jobs that might not require the volume applicable to closed chain of custody.

Glossary

action threshold: a level of incompatible plant pressure (e.g., species, density, height, location, or condition) where vegetation management control methods should occur to prevent conditions from reaching the tolerance levels.

active ROW: strip of land occupied by energized transmission facilities, excluding inactive or unused ROW portions set aside for other facilities or future construction.

adaptive management: a structured, iterative process of natural resource management that improves future management based on system monitoring.

allelopathy: ability of some plants to release chemicals that suppress other plant species around them.

ANSI A300: the *American National Standard for Tree Care Operations—Tree, Shrub, and Other Woody Plant Management—Standard Practices.* American national consensus standard series for tree care.

ANSI Z133: the *American National Standard for Arboricultural Operations—Safety Requirements.* American national consensus standard for arboricultural safety.

basal treatment: selective application of an oil-based herbicide mixture around the lower stem, root collar, and exposed roots of an incompatible plant.

best management practices (BMPs): best available, industry-recognized courses of action, in consideration of the benefits and limitations, based on scientific research, current knowledge, and applicable standards.

biological methods: management of vegetation by establishment and conservation of compatible plant communities using competition, allelopathy, animals, insects, or pathogens.

border zone: a section of an electric transmission or pipeline ROW that extends a specified distance from either side of the wire or pipe zone to the ROW edge, usually managed to promote mixed vegetation below a specified height (contrast with *pipe zone* and *wire zone*).

broadcast-foliar treatment: nonselective application of low concentrations of a herbicide mixture to the leaves of plants or groups of plants across a wide area.

chemical control methods: management of incompatible vegetation using herbicides or growth regulators.

chemical side pruning: selective application of a nonsystemic herbicide or growth regulator mixture to reduce or eliminate growth on treated portions of a plant.

closed chain of custody: an end-to-end process of documented ownership for herbicides, adjuvants, and containers from manufacturer through application, and the return of returnable, reusable containers to a customer blender for refilling and reuse.

compatible vegetation: plant forms that are consistent with the intended use of the site.

control methods[1]**:** procedures prescribed to attain the intended use of the site. Control methods include biological, chemical (herbicide and tree growth regulators), cultural, prescribed fire, and physical (manual and mechanical) methods.

cover-type conversion: method of control in which a desired plant community is achieved by providing compatible plants a competitive advantage over incompatible species.

cultural control methods: management of vegetation through the use of alternative land uses, including agricultural systems such as crops and pastures, parks, or other managed landscapes.

cut-stubble treatment: nonselective, broadcast application of a herbicide mixture to an area soon after physical control (e.g., mastication or mowing) by an applicator using a spray rig and calibrated equipment.

---

[1] ANSI A300, Part 7, refers to *treatment methods*, while this BMP uses the term *control methods*. The review committee differs from Part 7 out of deference to word usage common in IPM. The committee does not intend this departure to be interpreted as a recommendation to emphasize managing to control incompatible vegetation over managing to promote compatible vegetation.

cut-stump treatment: selective application of an oil-based herbicide mixture to the remaining stump, bark, root collar, and exposed roots.

cut-surface treatment: application of a herbicide mix, usually water based, to a freshly cut stump, concentrating on the sapwood and cambium.

debris: cut vegetation remaining after maintenance operations.

distribution lines: in an electric utility system, electric supply lines usually energized between 2.4 and 34.5 kV (contrast with *transmission lines*).

early successional plant communities: plant communities dominated by annual and perennial herbaceous plants that develop soon after disturbance.

early successional wildlife: animal species that use early successional plant communities as habitat.

easement: a document establishing the right to cross or otherwise use someone else's land for a specified purpose (see *prescriptive right*).

ecotone: an area where two plant communities integrate and transition from one to another.

effective border zone: a modification of the traditional border zone concept where a high-voltage line has sufficient ground-to-conductor clearance to accommodate mixed vegetation below a specified height throughout the ROW.

electrical fault: unintentional conducting path or blockage of a current in an electrical system.

exclusion zone: area in and around the base of power line structures that is maintained free of woody vegetation.

foliar application: selective or nonselective application of a herbicide mixture to the leaves of plants.

frilling: selective application of a herbicide mixture into cuts in the trunk. Also called hack-and-squirt.

goal: a desired result or purpose; in vegetation management, a strategic ambition for a program.

habitat: environment suitable for sustaining a population of a given organism.

hack-and-squirt: see *frilling*.

herbicide: a pesticide used to control plants to slow or suppress their growth by interfering with botanical pathways.

herptile: amphibian or reptile.

high-volume foliar: see *hydraulic foliar*.

hydraulic foliar: selective application of low concentrations of herbicide mixture across a relatively wide area on the leaves of individual plants.

incompatible vegetation: plant forms that are inconsistent with the intended use of a site.

integrated pest management (IPM): a method for managing pests that combines appropriate preventive and therapeutic tactics into a single management strategy.

integrated vegetation management (IVM): a system of managing plant communities in which compatible and incompatible vegetation are identified, action thresholds are considered, control methods are evaluated, and selected controls are implemented to achieve specific objectives (see note 1).

level 1 or limited visual tree risk assessment: a visual inspection from a specified perspective such as foot, vehicle, or aerial patrol of an individual tree or a population of trees to identify specified targets, conditions, or obvious defects.

level 2 or basic tree risk assessment: a detailed, ground-based visual inspection of an individual tree and its surrounding site.

light detection and ranging (lidar): a remote-sensing method that uses light in the form of a pulsed laser to generate precise, three-dimensional measurements.

line: a distribution or transmission electric facility including wire, poles, and attachments.

logs: woody stems greater than 6 in (15 cm) in diameter that result from tree or large branch removal.

low-volume foliar: selective application of a concentrated herbicide mixture to the leaves of plants using a low-pressure system, often from a backpack sprayer.

manual methods: management of vegetation using hand-operated tools such as handsaws and small power tools; a type of physical control.

mastication: a mechanical control technique where woody vegetation is cut, ground, or mulched by machinery mounted on a carrier (e.g., excavator tractor or skidder).

masticator: a large machine used to cut, grind, or mulch woody vegetation.

mechanical methods: management of vegetation using equipment, including those mounted with saws, masticators, mowers, or other devices; a type of physical control.

minimum vegetation clearance distance (MVCD): a calculated minimum distance between electrical conductors and vegetation to prevent sparkover, for various altitudes and operating voltages.

mode of action (MOA): the means by which a herbicide achieves an intended effect.

mower: a mechanical device designed to cut herbaceous plant material.

mowing: a mechanical control technique where herbaceous vegetation is cut.

National Electrical Safety Code®: a standard in the United States covering basic provisions for safeguarding persons from hazards resulting from installation, operation, or maintenance of conductors and equipment in electric supply stations, overhead and underground electric supply, and communication lines. It also contains work rules for construction, maintenance, and operations of electric supply, and communication lines and equipment.

nonselective management: management of all vegetation within a prescribed area without regard to species.

objective: something worked or striven for; in vegetation management, desired outcomes of a maintenance plan.

pesticide label: legally enforceable information provided about a pesticide, including written, printed, or graphics on or attached to the pesticide or device or any of its containers or wrappers.

physical methods: management of incompatible plants using manual and mechanical processes to remove, control, or alter target plants.

pipe zone: area of a utility pipeline ROW over the pipe and extending out both sides to a specified distance (see *border zone*).

plant pressure: an expression of risk created by incompatible plants growing in conflict with vegetation management objectives, typically expressed in terms of clearance, height, density, species, or other factors.

practitioner: person employed internally or contracted by the organization with responsibility over IVM.

prescribed fire: management of vegetation using a planned, controlled fire.

prescriptive right: right to cross or otherwise use someone else's land for a specified purpose obtained through continual use without permission for a period established by law (see *easement*).

reclamation: establishment or reestablishment of IVM objectives in areas not actively maintained.

right-of-use: legal authority to utilize property.

right-of-way (ROW) (pl. rights-of-way): a corridor of land used for a specific purpose, such as an electric transmission or pipeline.

risk: the combination of the likelihood of an event and the severity of the potential consequences. In the context of IVM, risk is the likelihood of trees, tree branches, or other vegetation falling onto or growing into managed facilities, causing damage to or interrupting services, combined with the severity of the potential consequences.

sag and sway: movement of energized supply lines in response to wind, temperature, or electrical load.

selective methods: control methods used to target incompatible plants while promoting compatible vegetation.

site of action (SOA): location in a plant cell where a herbicide disrupts growth and development.

species of concern: plant or animal species requiring priority for conservation.

specification: a document stating a detailed, measurable plan or proposal for provision of a product or service.

stakeholder: a person or group that has an interest in or is affected by an activity or decision. External stakeholders are outside an organization responsible for IVM. Internal stakeholders work within an organization responsible for IVM.

statement of work: document detailing intended outcomes of a project.

stocking: a measure of tree height and density relative to a fully occupied site.

sustainability: management principle based on creating and maintaining conditions for humans and nature to exist in productive balance.

threat: a vegetation condition likely to cause damage to a target of consequence at any moment.

tolerance level: the maximum allowable incompatible-plant pressure (e.g., species, density, height, location, or condition), without unacceptable consequences.

topographical zone: an electric ROW area, such as a canyon or valley, where ground-to-conductor clearance is sufficiently high to accommodate timber species without a threat of encroaching on tolerance levels.

transmission lines: in an electric utility system, electric supply lines used to connect generating stations to and between substations usually energized above 34.5 kV (contrast with *distribution lines*).

tree growth regulator (TGR): substance that reduces plant growth.

ultra-low-volume foliar: selective application of a herbicide mixture to the leaves of plants using specialized nozzles and a paraffin-based carrier that penetrates leaves and reduces active-ingredient drift and evaporation.

unmanned aerial system (UAS): an aircraft without an onboard human pilot (drone).

vegetation maintenance plan: a systematic approach to maintaining vegetation for intended project outcomes.

vegetation management program: strategic plans, policies, goals, specifications, and procedures for the administration of vegetation management activities.

vegetation manager: an individual engaged in the profession of vegetation management who, through appropriate experience, education, and related training, possesses the competence to provide for or supervise an integrated vegetation management program.

wetland: land where water saturation is the dominant factor determining the nature of soil development and the types of plant and animal communities living in and on it.

windrow: a line of cut and piled vegetation.

wire zone: the section of a utility transmission ROW under the wires and extending out both sides to a specified distance, usually managed to promote low-growing vegetation (contrast with *border zone*).

Selected References

American National Standards Institute. 2013. *American National Standard for Tree Care Operations—Tree, Shrub, and Other Woody Plant Management—Standard Practices (Root Management)* (ANSI A300, Part 8). Manchester (NH, USA): Tree Care Industry Association.

American National Standards Institute. 2017. *American National Standard for Arboricultural Operations—Safety Requirements* (ANSI Z133). Champaign (IL, USA): International Society of Arboriculture.

American National Standards Institute. 2017. *American National Standard—Tree, Shrub, and Other Woody Plant Management—Standard Practices (Pruning)* (ANSI A300, Part 1). Manchester (NH, USA): Tree Care Industry Association.

American National Standards Institute. 2017. *American National Standard—Tree, Shrub, and Other Woody Plant Management—Standard Practices (Tree Risk Assessment a. Tree Failure)* (ANSI A300, Part 9). Manchester (NH, USA): Tree Care Industry Association.

American National Standards Institute. 2017. *The Standard for Program Management* (ANSI/PMI 08-002-2017). 4th Ed. Newton Square (PA, USA): Program Management Institute.

American National Standards Institute. 2018. *American National Standard—Tree, Shrub, and Other Woody Plant Management—Standard Practices (Soil Management a. Assessment, b. Modification, c. Fertilization, and d. Drainage)* (ANSI A300, Part 2). Manchester (NH, USA): Tree Care Industry Association.

American National Standards Institute. 2018. *American National Standard—Tree, Shrub, and Other Woody Plant Management—Standard Practices (Integrated Vegetation Management)* (ANSI A300, Part 7). Manchester (NH, USA): Tree Care Industry Association.

Ansley J, Clayton M, Lyons RK, Hart CR. 2017. *Integrating Herbicides in Wildlife Habitat Management: A Technical Guide for Brush Management with Herbicides.* Indianapolis (IN, USA): Dow AgroSciences. [Accessed 2020 July 5]. https://static1.squarespace.com/static/50d37c2ce4b09ff030bc2f7b/t

/59f2269fe4966bac115312f3/1509041826771Integrating+Herbicides+Final+Oct+2017.pdf.

Appelt P, Gartman D. 2004. *Integrated Vegetation Management on Natural Gas Pipeline Rights-of-Way*. Presentation to the Environmental Concerns in Right-of-Way Management, 8th International Symposium; September 12–16, 2004; Saratoga Springs, NY.

Ballard BD, McLoughlin KT, Nowak CA. 2007. New Diagrams and Applications for the Wire-Border Zone Approach to Vegetation Management on Electric Transmission Line Rights-of-Way. *Arboriculture & Urban Forestry.* 33(6):435-439.

Baskerville G. 1985. Adaptive Management: Wood Availability and Habitat Availability. *The Forestry Chronical.* 61:171-175.

Bonneau J, Mucha S. 2019. Climate Change Adaptation Strategies in VM. In: Espinoza A, Pupa NG, editors. *Environmental Concerns in the Rights-of-Way Management: 12th International Symposium Proceedings.* Forest Lake (MN, USA): Utility Arborist Association. Atlanta, GA. p. 371-380. [Accessed 2020 August 4]. https://www.ROW.org/past-proceedings/.

Bonneville Power Administration. 2000. *Transmission System Vegetation Management Program: Final Environmental Impact Statement* (DOE/EIS). Portland (OR, USA): Bonneville Power Administration.

Bramble WC, Byrnes WR, Hutnik RJ. 1985. Effects of a Special Technique for Right-of-Way Maintenance on Deer Habitat. *Journal of Arboriculture.* 11(9):278-284.

Bramble WC, Yahner RH, Byrnes WR. 1992. Breeding-Bird Population Changes Following Right-of-Way Maintenance Treatments. *Journal of Arboriculture.* 18:23-32.

Bureau of Land Management. 2005. *Vegetation Treatment Using Herbicides on Bureau of Land Management Lands in 17 Western States: Programmatic Environmental Impact Statements* (DES 05-56). Washington (DC, USA): US Department of the Interior, Bureau of Land Management.

Chaney WR. 2005. *Growth Retardants: A Promising Tool for Managing Urban Trees* (Purdue Extension. FGR: 252-W). West Lafayette (IN, USA): Purdue University.

Cieslewicz S, Novembri R. 2004. *Utility Vegetation Management Final Report.* Federal Energy Regulatory Commission, United States Government. Commissioned to support the Federal Investigation of the August 14, 2003, Northeast Blackout. Washington (DC, USA): Federal Energy Regulatory Commission.

Cieslewicz S, Novembri R. 2004. *Utility Vegetation Management: Trends, Issues, and Practices.* Novato (CA, USA): CN Utility Consulting, LLC.

Coastal Change Analysis Program (C-CAP) Land Cover Classifications. Washington (DC, USA): National Oceanic and Atmospheric Administration. [Accessed 2021 January 6]. https://coast.noaa.gov/digitalcoast/training/ccap-land-cover-classifications.html.

Craul PJ. 1999. *Urban Soils: Applications and Practices*. New York (NY, USA): John Wiley and Sons.

Doran GT. 1981. There Is a S.M.A.R.T. Way to Write Management's Goals and Objectives. *Management Review.* 70(11):35-36.

Dunster JA, Smiley ET, Matheny N, Lilly S. 2017. *Tree Risk Assessment Manual.* 2nd Ed. Champaign (IL, USA): International Society of Arboriculture.

Edison Electric Institute, US Government. 2016. *Memorandum of Understanding on Vegetation Management for Powerline Rights-of-Way*. Washington (DC, USA): Edison Electric Institute.

Egler FE, Foote SR. 1975. *The Plight of the Rightofway Domain: Victim of Vandalism*. Mt. Kisco (NY, USA): Futura Media Services.

Electronic Code of Federal Regulations (CDR). Title 40: Protection of the Environment. [Updated 2021 May 13; accessed 2020 August 9]. https://www.ecfr.gov/cgi-bin/text-idx?SID=6b8e629cbac80e3d59550a7e90b32058&mc=true&node=pt40.37.1508&rgn=div5#se40.37.1508_118.

Environmental Protection Agency. 2006. Pesticide Environmental Stewardship Program. Washington (DC, USA): United States Environmental Protection Agency. [Accessed 2012 September 27]. http://www.epa.gov/pesp/.

Environmental Protection Agency. Undated. Learn About Sustainability. Washington (DC, USA): United States Environmental Protection Agency. [Accessed 2021 February 15]. https://www.epa.gov/sustainability/learn-about-sustainability#what.

Environmental Protection Agency. Undated. Why Are Wetlands Important? [Accessed 2021 March 18]. https://www.epa.gov/wetlands/why-are-wetlands-important.

Evans JA, Tranel P, Hager A, Schutte B, Wu C, Chatman LA, Davis AS. 2015. Managing the Evolution of Herbicide Resistance. *Pest Management Science.* 72:74-80.

Federal Energy Regulatory Commission. 2004. Utility Vegetation Management and Bulk Electric Reliability Report from the Federal Energy Regulatory Commission. Washington (DC, USA): Federal Energy Regulatory Commission, United States Government.

Fowler A, editor. 2015. Arkansas Wildlife Action Plan. Little Rock (AR, USA): Arkansas Game and Fish Commission.

Gilman EF, Lilly SJ, Smiley ET. 2019. *Best Management Practices: Tree Pruning.* 3rd Ed. Atlanta (GA, USA): International Society of Arboriculture.

Goodfellow JW. 1995. Engineering and Construction Alternatives to Line Clearance Tree Work. *Journal of Arboriculture.* 21(1):41-49.

Goodfellow JW. 2019. The Cost-Efficiency of IVM: A Comparison of Vegetation Management Strategies of Utility Rights-of-Way. BioCompliance Consulting, Inc. Utility Arborist Research Fund Grant #18-UARF-01.

Goodfellow JW. 2020. *Utility Tree Risk Assessment: Best Management Practices for Assessing the Risk of Tree Strikes to Electric Distribution and Transmission Lines and Related Equipment.* Atlanta (GA, USA): International Society of Arboriculture. Produced in collaboration with the Utility Arborist Association.

Goodfellow JW, Holt HA. 2011. *Utility Arborist Association Best Management Practices: Field Guide to Closed Chain of Custody for Herbicides in the Utility Vegetation Management Industry*. Champaign (IL, USA): International Society of Arboriculture.

Harold JM, Lowe ZF, Dukes JS. 2014. Integrated Vegetation Management (IVM) for INDOT Roadsides. West Lafayette (IN, USA): Purdue University. Report Number FHWA/IN/JTRP-2013/08. [Accessed 2020 August 20]. https://docs.lib.purdue.edu/jtrp/1550/.

Hutnik RJ, Bramble WC, Byrnes WR. 1987. Seedbed Contents on an Electric Transmission Right-of-Way. In: Byrnes WR, Holt HA, editors. Proceedings Fourth Symposium on Environmental Concerns in Right-of-Way Management. West Lafayette (IN, USA): Purdue University.

Institute of Electrical and Electronics Engineers. 2012. *National Electrical Safety Code*. New York (NY, USA): Institute of Electrical and Electronics Engineers.

International Fire Code Institute. 2018. *Urban-Wildland Interface Code*™. Whittier (CA, USA): International Fire Code Institute.

Johnstone RA, Haggie MR. 2012. Regional Vegetation Management Best Practices Case Studies: An Applied Approach for Utility and Wildlife Managers. In: Evans JM, Goodrich-Mahoney JW, Mutrie D, Reinemann J, editors. *Environmental Concerns in Rights-of-Way Management: 9th International Symposium*. Champaign (IL, USA): International Society of Arboriculture. p. 77-86.

Kempter G. 2004. *Best Management Practices: Utility Pruning of Trees.* Champaign (IL, USA): International Society of Arboriculture.

Krosby M, Tewksbury J, Haddad NM, Hoekstra J. 2010. Ecological Connectivity for a Changing Climate. *Conservation Biology.* 24(6):1686-1689.

Launchbaugh K, editor. 2006. *Targeted Grazing: A Natural Approach to Vegetation Management and Landscape Enhancement.* National Sheep Industry Improvement Center in cooperation with the American Sheep Industry Association. Centennial (CO, USA): Cottrell Printing.

Lembrechts JJ, Alexander JM, Cavieres LA, Haider S, Lenoir J, Kueffer C, McDougall K, Naylor BJ, Nunez MA, Pauchard A, Rew LJ, Nijs I, Malbau A. 2017. Mountain Roads Shift Native and Non-Native Plant Species' Ranges. *Ecography*. 40:353-364.

Lilly SJ. 2010. *Arborists' Certification Study Guide*. Champaign (IL, USA): International Society of Arboriculture.

Matheny NP, Clark JR. 1994. *A Photographic Guide to the Evaluation of Hazard Trees in Urban Areas*. 2nd Ed. Urbana (IL, USA): International Society of Arboriculture.

McGuire JL, Lawler JJ, McRae BH, Nuñez TA, Theobald DM. 2016. Achieving Climate Connectivity in a Fragmented Landscape. *Proceedings of the National Academy of Sciences*. 113(26):7195-7200.

McLoughlin KT. 2014. Integrated Vegetation Management: From Its Roots in IPM to the Present. In: Doucet GJ, editor. *Environmental Concerns in Rights-of-Way Management: 10th International Symposium*. Champaign (IL, USA): Utility Arborist Association.

Miller RH, Kempter G. 2018. *Utility Arboriculture: The Utility Specialist Certification Study Guide*. Champaign (IL, USA): International Society of Arboriculture.

Miller RW, Hauer RJ, Werner LP. 2015. *Urban Forestry: Planning and Managing Urban Greenspaces*. 3rd Ed. Long Grove (IL, USA): Waveland Press.

Miller TL, editor. 1993. *Oregon Pesticide Applicator Manual: A Guide to Safe Use and Handling of Pesticides*. Corvallis (OR, USA): Oregon State University.

New York State Department of Transportation. 2011. *Environmental Handbook for Transportation Operations: A Summary of the Environmental Requirements and Best Practices for Maintaining and Constructing Highways and Transportation Systems*. Albany (NY, USA): NYSDOT Environmental Science Bureau. [Accessed 2020 September 4]. https://www.dot.ny.gov/divisions/engineering/environmental-analysis/repository/oprhbook.pdf.

Niering WA. 1958. Principles of Sound Right-of-Way Vegetation Management. *Economic Botany*. 12(2):140-144.

Niering WA, Egler FE. 1953. A Shrub Community of *Viburnum lentago*: Stable for Twenty-Five Years. *Ecology*. 36(2):63-73.

North American Electric Reliability Council. 2008. Transmission Vegetation Management Standard FAC-003-2: Technical Reference. Washington (DC, USA): North American Electric Reliability Council.

Nowak CA, Ballard BD. 2005. A Framework for Applying Integrated Vegetation Management on Rights-of-Way. *Journal of Arboriculture*. 31(1):28-37.

OSHA, United States Department of Labor. Electric Power Generation, Transmission and Distribution Industry. [Accessed 2021 May 16]. https://www.osha.gov/power-generation/resources.

Ottoboni AM. 1997. *The Dose Makes the Poison*. 2nd Ed. New York (NY, USA): John Wiley and Sons.

Phillips-Mao L. 2017. Restoring Your Woody-Invaded Prairie to Conservation Prairie. Arlington (VA, USA): Nature Conservancy. [Accessed 2020 January 22]. https://www.nature.org/content/dam/tnc/nature/en/documents/Restoration-Guide-Woody-Invaded-to-Conservation-Prairie.pdf.

Richards NA. 1973. *Old Field Vegetation as an Inhibitor of Tree Vegetation*. New York (NY, USA): The Gary Arboretum of the New York Botanical Garden.

ROW Stewardship Council. 2016. Accreditation Standards for Assessing IVM Excellence. [Accessed 2020 January 3]. https://www.rowstewardship.org/standards.

Russell DB, Wischkaemper JW, Benner C. 2011. Best Practices in Vegetation Management for Enhancing Electric Service in Texas. College Station (TX, USA): Texas A&M University. Submitted to Public Utility Commission of Texas, Austin, TX. PUCT Project 38257.

Savitz AW. 2014. *The Triple Bottom Line*. San Francisco (CA, USA): John Wiley and Sons.

Shamoun SF. 1999. Application of Biological Control to Vegetation Management in Forestry. In: *Proceedings of the X International Symposium on Biological Control*; July 4–14, 1999; Montana State University, Bozeman, MT.

Shatford J, Hibbs D, Norris L. 2003. Identifying Plant Communities Resistant to Conifer Establishment Along Utility Rights-of-Way in Washington and Oregon, US. *Journal of Arboriculture*. 29(3):172-176.

Smiley TE, Matheny N, Lilly S. 2011. *Best Management Practices: Tree Risk Assessment*. Champaign (IL, USA): International Society of Arboriculture.

Sogge MK, Sferra SJ, Paxton EH. 2006. Tamarix as Habitat for Birds: Implications for Riparian Restoration in the Southwestern United States. *Restoration Ecology*. 16(1):146-154.

Stedman J, Brockbank R. 2012. Integrated Vegetation Management on Pipeline Rights-of-Way: Part One. *Utility Arborist Newsline*. 3(1):1-5.

United States Army Corps of Engineers. 2016. Construction Mat Best Management Practices (BMPs). United States Army Corps of Engineers, New England District. [Accessed 2021 March 13]. https://www.nae.usace.army.mil/Portals/74/docs/regulatory/StateGeneralPermits/MA/ConstructionMatBMPs.pdf.

United States Department of Agriculture. 2005. *Program for Biological Control of Saltcedar (*Tamarix *spp.) in Thirteen States: Environmental Assessment*. Fort Collins (CO, USA): Animal and Plant Health Inspection Service, US Department of Agriculture, Western Region. [Accessed 2020 January 10]. https://www.aphis.usda.gov/plant_health/ea/downloads/salteafonsi.pdf.

United States Department of Transportation. 2012. TECHBRIEF: Construction Quality Assurance for Design-Build Highway Projects. FHWA Publication No. FHWA-HRT-12-039. [Accessed 2020 March 3]. file:///C:/Users/Randym/OneDrive%20-%20Wright%20Service%20Corp/2020/Articles/Fed%20Highway%20Administration%20QA.pdf.

Vandenbroucke JE, Gaucher C, Major N. 2005. Biological Vegetation Management: An Alternative to Chemical Pesticides. *Journal of Arboriculture*. 31(5):251-256.

Weir JR. 2009. *Conducting Prescribed Fires: A Comprehensive Manual.* College Station (TX, USA): Texas A&M University Press.

Wildlife Habitat Council. Silver Spring (MD, USA): Wildlife Habitat Council. [Accessed 2020 July 8]. https://www.wildlifehc.org.

Yahner RH. 2004. Wildlife Response to More Than 50 Years of Vegetation Maintenance on Pennsylvania Right-of-Way. *Journal of Arboriculture.* 30(2):123-126.

Yahner RH, Bramble WC, Byrnes WR. 2001. Effect of Vegetation Maintenance of an Electric Transmission Right-of-Way on Reptile and Amphibian Populations. *Journal of Arboriculture.* 27:24-28.

Yahner RH, Hutnik RJ. 2004. Integrated Vegetation Management on an Electric Transmission Right-of-Way in Pennsylvania, US. *Journal of Arboriculture.* 30:295-300.

About the Author

Randall H. Miller is director of research and development at CNUC. He performs formal utility vegetation management industry survey studies and utility vegetation management program reviews among other responsibilities for CNUC. Prior to joining CNUC, he worked in PacifiCorp's vegetation management department for more than 23 years, including 18 as system forester and director of vegetation management.

Miller holds a BS in horticulture from the University of Wisconsin–Madison and an MS in urban forestry from the University of Wisconsin–Stevens Point. He is an ISA Board Certified Master Arborist® and an ISA Certified Arborist Utility Specialist® (IL-0225 BU) and TRAQ (Tree Risk Assessment Qualification) qualified. He is a past chair of the TREE Fund Board of Trustees, past president of the Utility Arborist Association, served twice as the chair of the Edison Electric Institute Vegetation Management Task Force, and serves on the International Society of Arboriculture Board of Directors. In 2005, he received the ISA Author's Citation, and he is widely published, with over 85 citations, most notably as coauthor with Geoff Kempter of *Utility Arboriculture: The Utility Specialist Certification Study Guide*.